Human Embyogenesis

Contents

Chapter 1

Human embryogenesis

This article is about Human embryogenesis. For Embryogenesis in general, see Embryogenesis.

Human embryogenesis is the process of cell division and

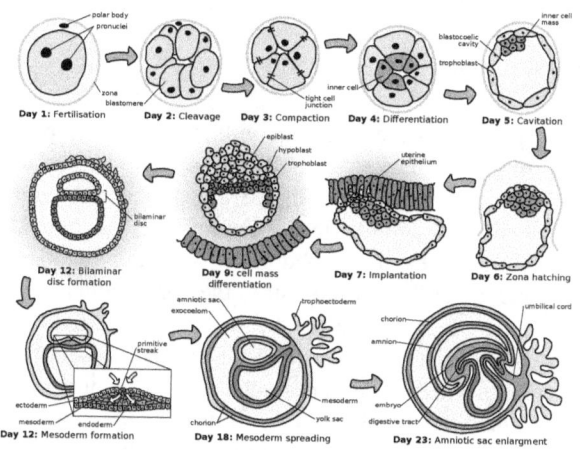

The initial stages of human embryogenesis.

cellular differentiation of the embryo that occurs during the early stages of development. In biological terms, human development entails growth from a one celled zygote to an adult human being. Fertilisation occurs when the sperm cell successfully enters and fuses with an egg cell (ovum). The genetic material of the sperm and egg then combine to form a single cell called a zygote and the germinal stage of prenatal development commences.[1] Embryogenesis covers the first eight weeks of development and at the beginning of the ninth week the embryo is termed a fetus. **Human embryology** is the study of this development during the first eight weeks after fertilisation. The normal period of gestation (pregnancy) is nine months or 38 weeks.

The germinal stage, refers to the time from fertilization, through the development of the early embryo until implantation is completed in the uterus. The germinal stage takes around 10 days.[2]

During this stage, the zygote, which is defined as an embryo because it contains a full complement of genetic material, begins to divide, in a process called cleavage. A blastocyst is then formed and implanted in the uterus. Embryogenesis continues with the next stage of gastrulation when the three germ layers of the embryo form in a process called histogenesis, and the processes of neurulation and organogenesis follow. The embryo is referred to as a fetus in the later stages of prenatal development, usually taken to be at the beginning of the ninth week. In comparison to the embryo, the fetus has more recognizable external features, and a more complete set of developing organs. The entire process of embryogenesis involves coordinated spatial and temporal changes in gene expression, cell growth and cellular differentiation. A nearly identical process occurs in other species, especially among chordates.

1.1 Germinal stage

1.1.1 Fertilization

Fertilization takes place when the spermatozoon has successfully entered the ovum and the two sets of genetic material carried by the gametes, fuse together, resulting in the zygote, (a single diploid cell). This usually takes place in the ampulla of one of the fallopian tubes. Successful fertilisation is enabled by three processes, which also act as controls to ensure species-specificity. The first is that of chemotaxis which directs the movement of the sperm towards the ovum. Secondly there is an adhesive compatibility between the sperm and the egg. With the sperm adhered to the ovum, the third process of acrosomal reaction takes place; the front part of the spermatozoon head is capped by an acrosome which contains digestive enzymes to break down the zona pellucida and allow its entry.[3] The entry of the sperm causes calcium to be released which blocks entry to other sperm cells. A parallel reaction takes place in the ovum called the zona reaction. This sees the release of cortical granules that release enzymes which digest sperm receptor proteins, thus preventing polyspermy. The granules also fuse with the plasma membrane and modify the

1

zona pellucida in such a way as to prevent further sperm entry.

The zygote contains the combined genetic material carried by both the male and female gametes which consists of the 23 chromosomes from the nucleus of the ovum and the 23 chromosomes from the nucleus of the sperm. The 46 chromosomes undergo changes prior to the mitotic division which leads to the formation of the embryo having two cells.

1.1.2 Cleavage stage

Further information: Cleavage (embryo)
This first division marks the beginning of the cleavage

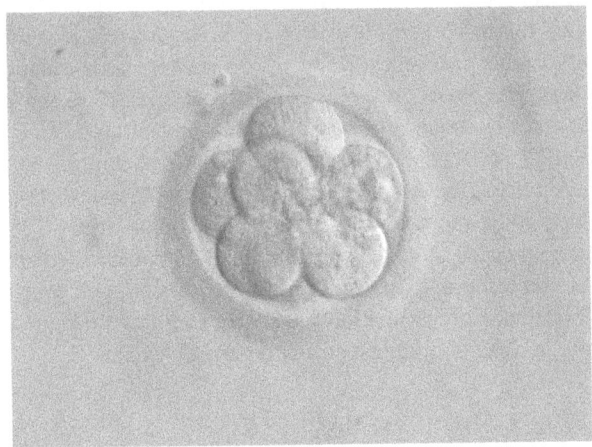

8-cell embryo, at 3 days

process which continues with the division of the first two cells by mitosis to give four cells which then divide to give eight cells and so on. This is quite a slow process taking from 12 to 24 hours for each division. The dividing cells which are termed blastomeres (*blastos* Greek for sprout) are still enclosed within the strong membrane of glycoproteins (termed the zona pellucida) of the ovum, which the successful spermatozoon managed to penetrate. The zygote (which is large compared to any other cell) undergoes further cleavage, increasing the number of cells without any increase in the size of the initial zygote. This means that the proportion of nuclear genetic material is greater than that of the cytoplasm in each cell. When eight blastomeres have formed they are undifferentiated and aggregated into a sphere. The cells begin to form gap junctions by this time, enabling them to develop in an integrated way and co-ordinate their response to physiological signals and environmental cues.[4]

When the cells number about sixteen or thirty-two the solid sphere of cells is termed a morula.[5] At this stage the cells start to bind firmly together in a process called **compaction**, and cleavage continues as cellular differentiation begins.

1.1.3 Blastulation

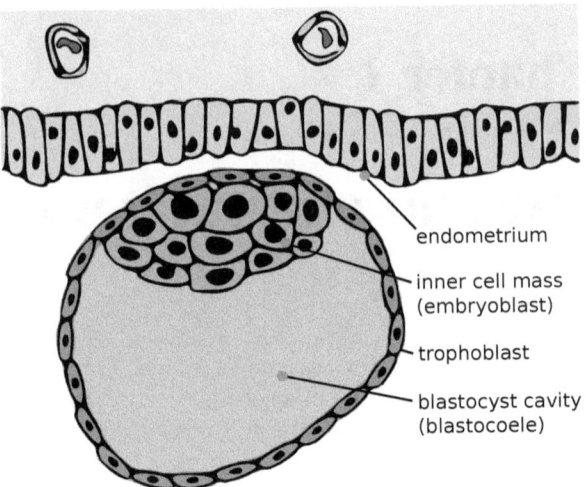

endometrium

inner cell mass (embryoblast)

trophoblast

blastocyst cavity (blastocoele)

Blastocyst with an inner cell mass and trophoblast.

Cleavage itself is the first stage in blastulation, the process of forming the blastocyst. Cells differentiate into an outer layer of cells (collectively called the trophoblast) and an inner cell mass. With further compaction the individual outer blastomeres, the trophoblasts, become indistinguishable, and are still enclosed within the zona pellucida. This compaction serves to make the structure watertight since the cells will later secrete fluid. The inner mass of cells differentiate to become embryoblasts and polarise at one end. They close together and form gap junctions in order to facilitate cellular communication. This polarisation leaves a cavity, the blastocoel in which is now termed the blastocyst. (In animals other than mammals, this is called the blastula). The trophoblasts secrete fluid into the blastocoel. By this time the size of the blastocyst has increased which makes it 'hatch' through the zone pellucida which then disintegrates.[6][7]

The inner cell mass will give rise to the embryo proper, the amnion, yolk sac and allantois, while the fetal part of the placenta will form from the outer trophoblast layer. The embryo plus its membranes is called the conceptus and by this stage the conceptus is in the uterus. The zona pellucida ultimately disappears completely, and the now exposed cells of the trophoblast allow the blastocyst to attach itself to the endometrium, where it will implant. The formation of the hypoblast and epiblast occurs at the beginning of the second week, which are the two main layers of the bilaminar germ disc.[8] Either the inner cells embryoblast or the outer cells trophoblast will turn into two sub layers each other.[9] The inner cells will turn into the hypoblast layer that will surround the other layer called epiblast layer, and these layers will form the embryonic disc in which the embryo will develop.[8][9] The place where the embryo develops is called the amniotic cavity, which is the inside the disc.[8] Also the trophoblast will develop two sub-layers; the cytotrophoblast that is front of the syncytiotrophoblast that is inside of the endometrium.[8] Next, another layer called the exocoelomic membrane or Heuser's membrane will appear and surround the cytotrophoblast, as well as the primitive yolk sac.[9] The syncytiotrophoblast will grow and will enter a phase called lacunar stage, in which some vacuoles will appear and be filled by blood in the following days.[8][9] The development of the yolk sac starts with the hypoblastic flat cells that form the exocoelomic membrane, which will coat the inner part of the cytotrophoblast to form the primitive yolk sac. An erosion of the endothelial lining of the maternal capillaries by the syncytiotrophoblastic cells of the sinusoids will form where the blood will begin to penetrate and flow through the trophoblast to give rise to the uteroplacental circulation.[10][11] Subsequently new cells derived from yolk sac will be established between trophoblast and exocelomic membrane and will give rise to extra-embryonic mesoderm, which will form cavities known as chorionic cavity.[9]

At the end of the second week of development, some cells of the trophoblast penetrate and form rounded columns into the syncytiotrophoblast. These columns are known as primary villi. At the same time, other migrating cells form into the exocelomic cavity, a new cavity named as secondary or definitive yolk, smaller in size than the primitive yolk sac.[9][10]

1.1.4 Implantation

Main article: Implantation (human embryo)
After ovulation, the endometrial lining becomes trans-

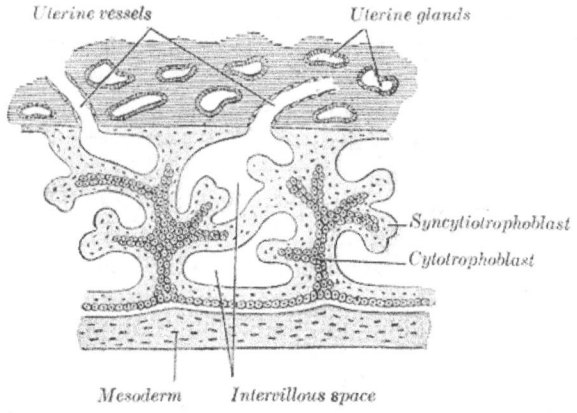

Trophoblast differentiation

formed into a secretory lining in preparation of accepting the embryo. It becomes thickened with its secretory glands becoming elongated, and is increasingly vascular. This lining of the uterine cavity (or womb), is now known

as the decidua and it produces a great number of large decidual cells in its increased interglandular tissue. The trophoblast then differentiates into an inner layer, the cytotrophoblast and an outer layer, the syncytiotrophoblast. The cytotrophoblast contains cuboidal epithelial cells having cell boundaries and are the source of dividing cells and the syncytiotrophoblast is a layer without cell boundaries.

The syncytiotrophoblast implants the blastocyst in the decidual epithelium, by projections of chorionic villi forming the embryonic part of the placenta. The placenta develops once the blastocyst is implanted, and forms to connect the embryo to the uterine wall. The decidua here is termed the decidua basalis and lies between the blastocyst and the myometrium and forms the maternal part of the placenta. The implantation is assisted by hydrolytic enzymes that erode the epithelium. The syncytiotrophoblast also produces human chorionic gonadotropin (hCG), a hormone that stimulates the release of progesterone from the corpus luteum. Progesterone enriches the uterus with a thick lining of blood vessels and capillaries so that it can sustain the developing embryo. The villi begin to branch and contain blood vessels of the embryo. Other villi, called terminal or free villi, have the role of nutrient exchange. The embryo is joined to the trophoblastic shell by a narrow connecting stalk that develops into the umbilical cord to attach the placenta to the embryo.[9][12] Arteries in the decidua are remodelled to increase the maternal blood flow into the intervillous spaces of the placenta, allowing gas exchange to take place as well as the transfer of nutrients to the embryo. Waste products from the embryo will diffuse across the placenta.

As the syncytiotrophoblast starts to penetrate the uterine wall, the inner cell mass (embryoblast) also develops. The inner cell mass is the source of embryonic stem cells, which are pluripotent and can develop into any one of the three germ layer cells.

1.1.5 Embryonic disc

The embryoblast forms an embryonic disc which is a bilaminar disc of two layers, an upper layer the *epiblast* (primitive ectoderm), and a lower layer the *hypoblast* (primitive endoderm). The disc is stretched between what will become the amniotic cavity and the yolk sac. The epiblast is adjacent to the trophoblast and made of columnar cells; the hypoblast is closest to the blastocyst cavity, and made of cuboidal cells. The epiblast migrates away from the trophoblast downwards, forming the amniotic cavity, the lining of which is formed from *amnioblasts* developed from the epiblast. The hypoblast is pushed down and forms the yolk sac (exocoelomic cavity) lining. Some hypoblast cells migrate along the inner cytotrophoblast lining of the blastocoel, secreting an extracellular matrix along the way. These hypoblast cells and extracellular matrix are called *Heuser's membrane* (or *exocoelomic membrane*), and they cover the blastocoel to form the yolk sac (or *exocoelomic cavity*). Cells of the epiblast migrate along the outer edges of this reticulum and form the *extraembryonic mesoderm*, which makes it difficult to maintain the extraembryonic reticulum. Soon pockets form in the reticulum, which ultimately coalesce to form the *chorionic cavity* or *extraembryonic coelom*.

1.2   Gastrulation

Main article: Gastrulation

The **primitive streak**, a linear band of cells formed by the

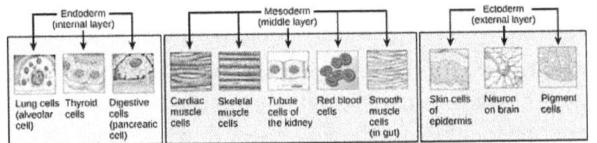

Histogenesis of the three germ layers

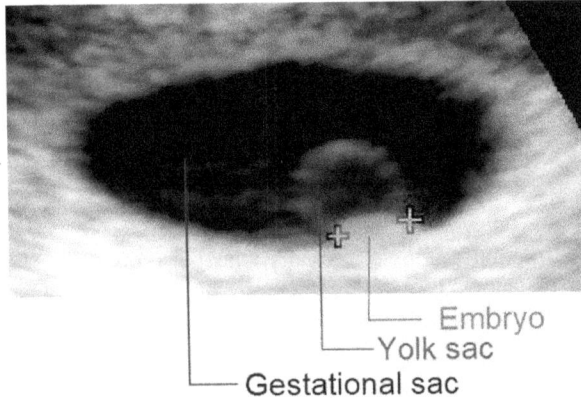

Artificially colored - gestational sac, yolk sac and embryo (measuring 3 mm at 5 weeks)

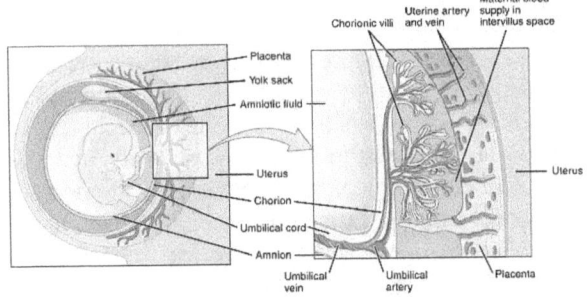

Embryo attached to placenta in amniotic cavity

migrating epiblast, appears, and this marks the beginning of **gastrulation**, which takes place around the sixteenth day (week 3) after fertilisation. The process of gastrulation reorganises the two-layer embryo into a three-layer embryo, and also gives the embryo its specific head-to-tail, and front-to-back orientation, by way of the primitive streak which establishes bilateral symmetry. A primitive node (or primitive knot) forms in front of the primitive streak which is the organiser of neurulation. A primitive pit forms as a depression in the centre of the primitive node which connects to the notochord which lies directly underneath. The node has arisen from epiblasts of the amniotic cavity floor, and it is this node that induces the formation of the neural plate which serves as the basis for the nervous system. The neural plate will form opposite the primitive streak from ectodermal tissue which thickens and flattens into the neural plate. The epiblast in that region moves down into the streak at the location of the primitive pit where the process called ingression, which leads to the formation of the mesoderm takes place. This ingression sees the cells from the epiblast move into the primitive streak in an epithelial-mesenchymal transition; epithelial cells become mesenchymal stem cells, multipotent stromal cells that can differentiate into various cell types. The hypoblast is pushed out of the way and goes on to form the amnion.The epiblast keeps moving and forms a second layer, the mesoderm. The epiblast has now differentiated into the three germ layers of the embryo, so that the bilaminar disc is now a trilaminar disc, the gastrula.

The three germ layers are the ectoderm, mesoderm and endoderm, and are formed as three overlapping flat discs. It is from these three layers that all the structures and organs of the body will be derived through the processes of somitogenesis, histogenesis and organogenesis.[13] The embryonic endoderm is formed by invagination of epiblastic cells that migrate to the hypoblast, while the mesoderm is formed by the cells that develop between the epiblast and endoderm. In general, all germ layers will derive from the epiblast.[9][12] The upper layer of ectoderm will give rise to the outermost layer of skin, central and peripheral nervous systems, eyes, inner ear, and many connective tissues.[14] The middle layer of mesoderm will give rise to the heart and the beginning of the circulatory system as well as the bones, muscles and kidneys. The inner layer of endoderm will serve as the starting point for the development of the lungs, intestine and bladder.

Following ingression, a blastopore develops where the cells have ingressed, in one side of the embryo and it deepens to become the archenteron, the first formative stage of the gut. The blastopore becomes the anus whilst the gut tunnels through the embryo to the other side where the opening becomes the mouth. With a functioning digestive tube, gastrulation is now completed and the next stage of neurulation can begin.

1.3 Neurulation

Main article: Neurulation
Following gastrulation, the ectoderm gives rise to epithe-

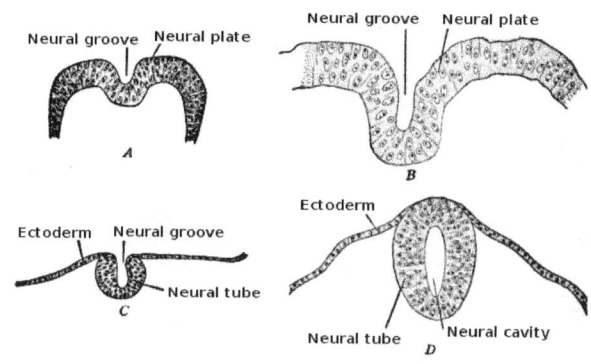

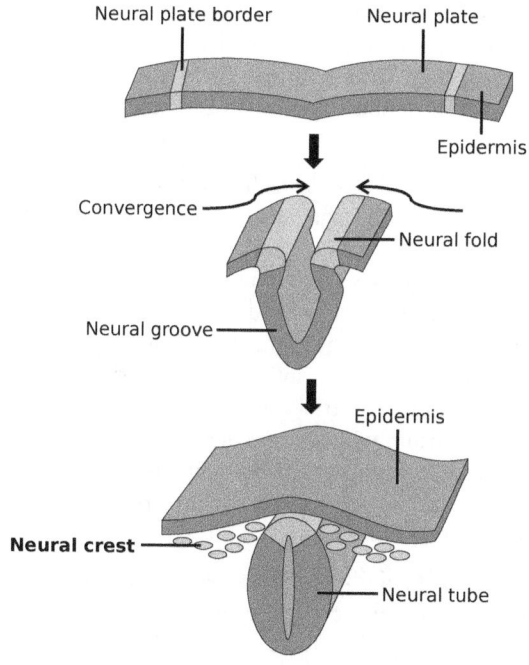

Neural plate

lial and neural tissue, and the gastrula is now referred to as the neurula. The neural plate that has formed as a thickened plate from the ectoderm, continues to broaden and its ends start to fold upwards as neural folds. Neurulation refers to this folding process whereby the neural plate is transformed into the neural tube, and this takes place during the fourth week. They fold, along a shallow neural groove which has formed as a dividing median line in the neural plate. This deepens as the folds continue to gain height, when they will meet and close together. The cells that mi-

grate through the most cranial part of the primitive line form the paraxial mesoderm, which will give rise to the somitomeres that in the process of **somitogenesis** will differentiate into somites that will form the sclerotome, the syndetome,[15] the myotome and the dermatome to form cartilage and bone, tendons, dermis (skin), and muscle. The intermediate mesoderm gives rise to the urogenital tract and consists of cells that migrate from the middle region of the primitive line. Other cells migrate through the caudal part of the primitive line and form the lateral mesoderm, and those cells migrating by the most caudal part contribute to the extraembryonic mesoderm.[9][12]

The embryonic disc begins flat and round, but eventually elongates to have a wider cephalic part and narrow-shaped caudal end.[8] At the beginning, the primitive line extends in cephalic direction and 18 days after fertilization returns caudally until it disappears. In the cephalic portion, the germ layer shows specific differentiation at the beginning of the 4th week, while in the caudal portion it occurs at the end of the 4th week.[9] Cranial and caudal neuropores become progressively smaller until they close completely (by day 26) forming the neural tube.[16]

1.3.1 Development of the nervous system

Main article: Neural development
Late in the fourth week, the superior part of the neu-

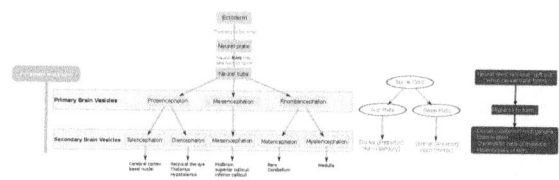

ral tube flexes at the level of the future midbrain—the mesencephalon. Above the mesencephalon is the prosencephalon (future forebrain) and beneath it is the rhombencephalon (future hindbrain).

The optical vesicle (which eventually becomes the optic nerve, retina and iris) forms at the basal plate of the prosencephalon. The alar plate of the prosencephalon expands to form the cerebral hemispheres (the telencephalon) whilst its basal plate becomes the diencephalon. Finally, the optic vesicle grows to form an optic outgrowth.

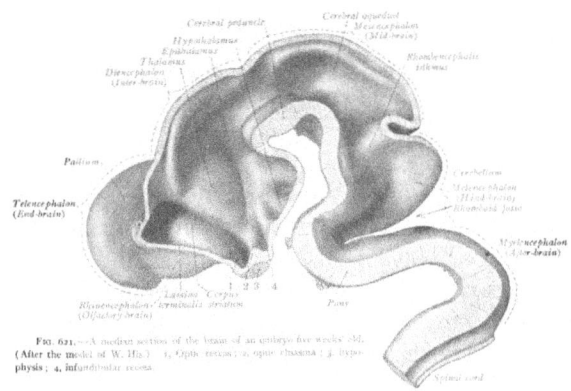

Spinal cord at 5 weeks

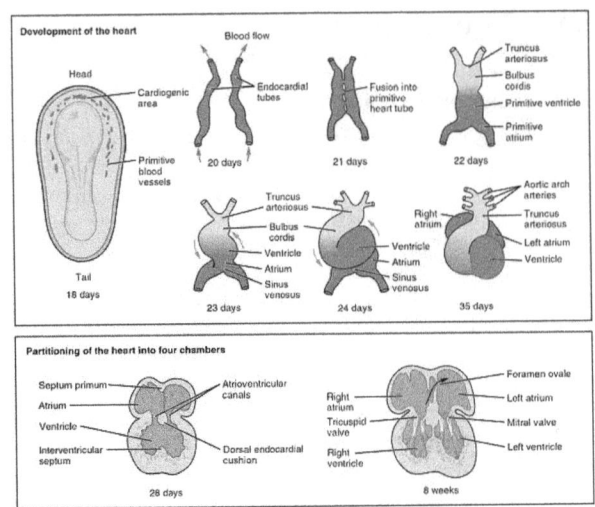

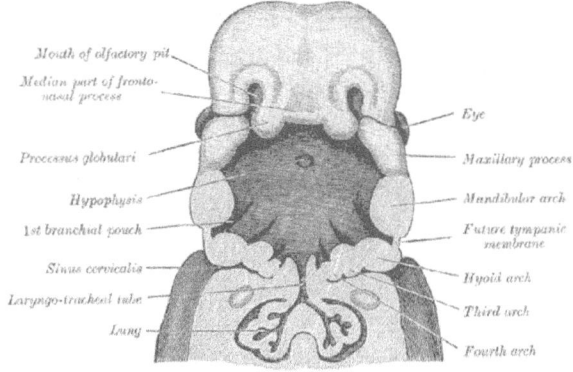

Head and neck at 32 days

1.4 Development of the heart and circulatory system

Main article: Heart development

The heart is the first functional organ to develop and starts to beat and pump blood at around 21 or 22 days.[17] Cardiac myoblasts and blood islands in the splanchnopleuric mesenchyme on each side of the neural plate, give rise to the cardiogenic region.[9]:165 This is a horseshoe-shaped area near to the head of the embryo. By day 19, following cell signalling, two strands begin to form as tubes in this region, as a lumen develops within them. These two endocardial tubes grow and by day 21 have migrated towards each other and fused to form a single primitive heart tube, the tubular heart. This is enabled by the folding of the embryo which pushes the tubes into the thoracic cavity.[18]

Also at the same time that the tubes are forming, vasculogenesis (the development of the circulatory system) has begun. This starts on day 18 with cells in the splanchnopleuric mesoderm differentiating into angioblasts that develop into flattened endothelial cells. These join to form small vesicles called angiocysts which join up to form long vessels called angioblastic cords. These cords develop into a pervasive network of plexuses in the formation of the vascular network. This network grows by the additional budding and sprouting of new vessels in the process of angiogenesis.[18]

The tubular heart quickly forms five distinct regions. From head to tail, these are the infundibulum, bulbus cordis, primitive ventricle, primitive atrium, and the sinus venosus. Initially, all venous blood flows into the sinus venosus, and is propelled from tail to head to the truncus arteriosus. This will divide to form the aorta and pulmonary artery; the bulbus cordis will develop into the right (primitive) ventricle; the primitive ventricle will form the left ventricle; the primitive atrium will become the front parts of the left and right atria and their appendages, and the sinus venosus will develop into the posterior part of the right atrium, the sinoatrial node and the coronary sinus.[17]

Cardiac looping begins to shape the heart in a process called morphogenesis and this completes by the end of the fourth week. Programmed cell death (apoptosis) is involved in this process, at the joining surfaces enabling fusion to take place.[18] In the middle of the fourth week, the sinus venosus receives blood from the three major veins: the vitelline, the umbilical and the common cardinal veins.

During the first two months of development, the interatrial septum begins to form. This septum divides the primitive atrium into a right and a left atrium. Firstly it starts as a crescent-shaped piece of tissue which grows downwards as the septum primum. The crescent shape prevents the complete closure of the atria allowing blood to be shunted from the right to the left atrium through the opening known as the ostium primum. This closes with further development of the system but before it does, a second opening (the ostium secundum) begins to form in the upper atrium enabling the continued shunting of blood.[18]

A second septum (the septum secundum) begins to form to the right of the septum primum. This also leaves a small opening, the foramen ovale which is continuous with the previous opening of the ostium secundum. The septum primum is reduced to a small flap that acts as the valve of the foramen ovale and this remains until its closure at birth. Between the ventricles the septum inferius also forms which develops into the muscular interventricular septum.[18]

1.5 Clinical significance

Toxic exposures during the germinal stage may cause prenatal death resulting in a miscarriage, but do not cause developmental defects. However, toxic exposures in the embryonic period can be the cause of major congenital malformations, since the precursors of the major organ systems are now developing.

Each cell of the preimplantation embryo has the potential to form all of the different cell types in the developing embryo. This cell potency means that some cells can be removed from the preimplantation embryo and the remaining cells will compensate for their absence. This has allowed the development of a technique known as preimplantation genetic diagnosis, whereby a small number of cells from the preimplantation embryo created by IVF, can be removed by biopsy and subjected to genetic diagnosis. This allows embryos that are not affected by defined genetic diseases to be selected and then transferred to the mother's uterus.

Sacrococcygeal teratomas, tumours formed from different types of tissue, that can form, are thought to be related to primitive streak remnants, which ordinarily disappear.[8][9][11]

Spina bifida a congenital disorder is the result of the incomplete closure of the neural tube.

Vertically transmitted infections can be passed from the mother to the unborn child at any stage of its development.

Hypoxia a condition of inadequate oxygen supply can be a serious consequence of a preterm or premature birth.

1.6 See also

- CDX2
- Developmental biology
- Embryomics
- Eye development
- Gonadogenesis
- Limb development
- Potential person
- Recapitulation theory

1.7 References

[1] Sherk, Stephanie Dionne. "http://www.healthline.com/galecontent/prenatal-development". *Gale Encyclopedia of Children's Health, 2006*. Gale. Retrieved 6 October 2013. External link in |title= (help)

[2] "germinal stage". *Mosby's Medical Dictionary, 8th edition*. Elsevier. Retrieved 6 October 2013.

[3] "acrosome definition - Dictionary - MSN Encarta". Archived from the original on 2009-10-31. Retrieved 2007-08-15.

[4] Brison, D. R.; Sturmey, R. G.; Leese, H. J. (2014). "Metabolic heterogeneity during preimplantation development: the missing link?". *Human Reproduction Update* **20** (5): 632–640. doi:10.1093/humupd/dmu018. ISSN 1355-4786.

[5] Boklage, Charles E. (2009). *How New Humans Are Made: Cells and Embryos, Twins and Chimeras, Left and Right, Mind/Self/Soul, Sex, and Schizophrenia*. World Scientific. p. 217. ISBN 978-981-283-513-0.

[6] http://www.vanat.cvm.umn.edu/TFFLectPDFs/LectEarlyEmbryo

[7] Forgács, G. & Newman, Stuart A. (2005). "Cleavage and blastula formation". *Biological physics of the developing embryo*. Cambridge University Press. p. 27. ISBN 978-0-521-78337-8.

[8] Carlson, Bruce M. (1999) [1t. Pub. 1997]. "Chapter 4: Formation of germ layers and initial derivatives". *Human Embryology & Developmental Biology*. Mosby, Inc. pp. 62–68. ISBN 0-8151-1458-3.

[9] Sadler, T.W.; Langman, Jan (2012) [1st. Pub. 2001]. "Chapter 3: Primera semana del desarrollo: de la ovulación a la implantación". In Seigafuse, sonya. *Langman, Embriología médica*. Lippincott Williams & Wilkins, Wolters Kluwer. pp. 29–42. ISBN 978-84-15419-83-9.

[10] Moore, Keith L.; Persaud, V.N. (2003) [1t. Pub. 1996]. "Chapter 3: Formation of the bilaminar embryonic disc: second week". *The Developing Human, Clinically Oriented Embryology*. W B Saunders Co. pp. 47–51. ISBN 0-7216-9412-8.

[11] Larsen, William J.; Sherman, Lawrence S.; Potter, S. Steven; Scott, William J. (2001) [1t. Pub. 1998]. "Chapter 2: Bilaminar embryonic disc development and establishment of the uteroplacental circulation". *Human Embryology*. Churchill Livingstone. pp. 37–45. ISBN 0-443-06583-7.

[12] Smith Agreda, Víctor; Ferrés Torres, Elvira; Montesinos Castro-Girona, Manuel (1992). "Chapter 5: Organización del desarrollo: Fase de germinación". *Manual de embriología y anatomía general*. Universitat de València. pp. 72–85. ISBN 84-370-1006-3.

[13] Ross, Lawrence M. & Lamperti, Edward D., ed. (2006). "Human Ontogeny: Gastrulation, Neurulation, and Somite Formation". Atlas of anatomy: general anatomy and musculoskeletal system. Thieme. ISBN 978-3-13-142081-7.lurl=https://books.google.com/books?id=NK9TgTaGt6UC&pg=PA6

[14] "Pregnancy week by week". Retrieved 28 July 2010.

[15] Brent AE, Schweitzer R, Tabin CJ (April 2003). "A somitic compartment of tendon progenitors". *Cell* **113** (2): 235–48. doi:10.1016/S0092-8674(03)00268-X. PMID 12705871. Retrieved 2014-04-20.

[16] Larsen, W J (2001). *Human Embryology* (3rd ed.). Elsevier. p. 87. ISBN 0-443-06583-7.

[17] Betts, J. Gordon (2013). *Anatomy & physiology*. pp. 787–846. ISBN 1938168135.

[18] Larsen, W J (2001). *Human Embryology* (3rd ed.). Elsevier. pp. 170–190. ISBN 0-443-06583-7.

1.8 External links

- Photo of blastocyst in utero

- Slideshow: In the Womb

- Online course in embryology for medicine students developed by the universities of Fribourg, Lausanne and Bern

Chapter 2

Human fertilization

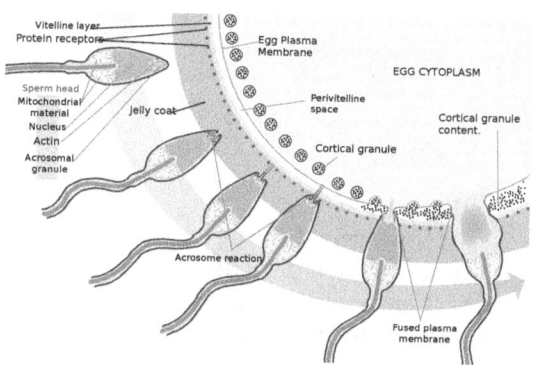

The acrosome reaction for a sea urchin, a similar process. Note that the picture shows several stages of one and the same spermatozoon - only one penetrates the ovum

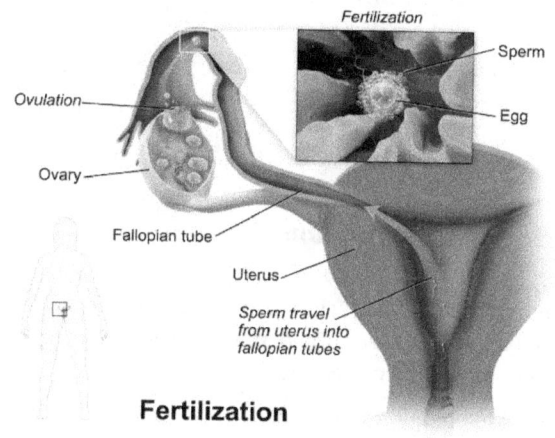

Illustration depicting ovulation and fertilization.

Human fertilization is the union of a human egg and sperm, usually occurring in the ampulla of the fallopian tube. The result of this union is the production of a zygote cell, or fertilized egg, initiating prenatal development. Scientists discovered the dynamics of human fertilization in the nineteenth century.[1]

The process of fertilization involves a sperm fusing with an ovum. The most common sequence begins with ejaculation during copulation, follows with ovulation, and finishes with fertilization. Various exceptions to this sequence are possible, including artificial insemination, *in vitro* fertilization, external ejaculation without copulation, or copulation shortly after ovulation.[2][3][4] Upon encountering the secondary oocyte, the acrosome of the sperm produces enzymes which allow it to burrow through the outer jelly coat of the egg. The sperm plasma then fuses with the egg's plasma membrane, the sperm head disconnects from its flagellum and the egg travels down the Fallopian tube to reach the uterus.

In vitro fertilization (IVF) is a process by which egg cells are fertilized by sperm outside the womb, *in vitro*.

2.1 Anatomy

2.1.1 Corona radiata

The sperm bind through the corona radiata, a layer of follicle cells on the outside of the secondary oocyte. Fertilization occurs when the nucleus of both a sperm and an egg fuse to form a diploid cell, known as zygote. The successful fusion of gametes forms a new organism.

2.1.2 Cone of attraction and perivitelline membrane

Where the spermatozoon is about to pierce, the yolk (ooplasm) is drawn out into a conical elevation, termed the cone of attraction or reception cone. Once the spermatozoon has entered, the peripheral portion of the yolk changes into a membrane, the perivitelline membrane, which prevents the passage of additional spermatozoa.[5]

2.1.3 Sperm preparation

Further information: Acrosome reaction

At the beginning of the process, the sperm undergoes a series of changes, as freshly ejaculated sperm is unable or poorly able to fertilize.[6] The sperm must undergo capacitation in the female's reproductive tract over several hours, which increases its motility and destabilizes its membrane, preparing it for the acrosome reaction, the enzymatic penetration of the egg's tough membrane, the zona pellucida, which surrounds the oocyte.

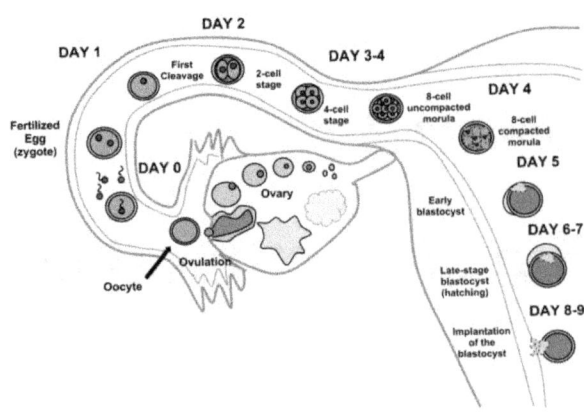

Fertilization and implantation in humans.

2.1.4 Zona pellucida

After binding to the corona radiata the sperm reaches the zona pellucida, which is an extra-cellular matrix of glycoproteins. A special complementary molecule on the surface of the sperm head binds to a ZP3 glycoprotein in the zona pellucida. This binding triggers the acrosome to burst, releasing enzymes that help the sperm get through the zona pellucida.

Some sperm cells consume their acrosome prematurely on the surface of the egg cell, facilitating the penetration by other sperm cells. As a population, sperm cells have on average 50% genome similarity so the premature acrosomal reactions aid fertilization by a member of the same cohort.[7] It may be regarded as a mechanism of kin selection.

Recent studies have shown that the egg is not passive during this process.[8][9]

Cortical reaction

Once the sperm cells find their way past the zona pellucida, the cortical reaction occurs. Cortical granules inside the secondary oocyte fuse with the plasma membrane of the cell, causing enzymes inside these granules to be expelled by exocytosis to the zona pellucida. This in turn causes the glyco-proteins in the zona pellucida to cross-link with each other — i.e. the enzymes cause the ZP2 to hydrolyse into ZP2f — making the whole matrix hard and impermeable to sperm. This prevents fertilization of an egg by more than one sperm. The cortical reaction and acrosome reaction are both essential to ensure that only one sperm will fertilize an egg.[10]

2.2 Fusion

After the sperm enters the cytoplasm of the oocyte (also called ovocyte), the cortical reaction takes place, preventing other sperm from fertilizing the same egg. The oocyte now undergoes its second meiotic division producing the haploid ovum and releasing a polar body. The sperm nucleus then fuses with the ovum, enabling fusion of their genetic material.

2.2.1 Cell membranes

The cell membranes of the secondary oocyte and sperm fuse.

2.2.2 Transformations

In preparation for the fusion of their genetic material both the oocyte and the sperm undergo transformations as a reaction to the fusion of cell membranes.

The oocyte completes its second meiotic division. This results in a mature ovum. The nucleus of the oocyte is called a pronucleus in this process, to distinguish it from the nuclei that are the result of fertilization.

The sperm's tail and mitochondria degenerate with the formation of the male pronucleus. This is why all mitochondria in humans are of maternal origin. Still, a considerable amount of RNA from the sperm is delivered to the resulting embryo and likely influences embryo development and the phenotype of the offspring.[11]

2.2.3 Replication

The pronuclei migrate toward the center of the oocyte, rapidly replicating their DNA as they do so to prepare the zygote for its first mitotic division.[12]

2.2.4 Mitosis

The male and female pronuclei don't fuse, although their genetic material do. Instead, their membranes dissolve, leaving no barriers between the male and female chromosomes. During this dissolution, a mitotic spindle forms between them. The spindle captures the chromosomes before they disperse in the egg cytoplasm. Upon subsequently undergoing mitosis (which includes pulling of chromatids towards centrioles in anaphase) the cell gathers genetic material from the male and female together. Thus, the first mitosis of the union of sperm and oocyte is the actual fusion of their chromosomes.[12]

Each of the two daughter cells resulting from that mitosis has one replica of each chromatid that was replicated in the previous stage. Thus, they are genetically identical.

2.3 Fertilization age

Fertilization is the event most commonly used to mark the zero point in descriptions of prenatal development of the embryo or fetus. The resultant age is known as *fertilization age, fertilizational age, embryonic age, fetal age* or *(intrauterine) developmental (IUD)[13] age.*

Gestational age, in contrast, takes the beginning of the last menstrual period (LMP) as the zero point. By convention, gestational age is calculated by adding 14 days to fertilization age and vice versa.[14] In fact, however, fertilization usually occurs within a day of ovulation, which, in turn, occurs on average 14.6 days after the beginning of the preceding menstruation (LMP).[15] There is also considerable variability in this interval, with a 95% prediction interval of the ovulation of 9 to 20 days after menstruation even for an average woman who has a mean LMP-to-ovulation time of 14.6.[16] In a reference group representing all women, the 95% prediction interval of the LMP-to-ovulation is 8.2 to 20.5 days.[15]

Fertilization age is sometimes used postnatally (after birth) as well to estimate various risk factors. For example, it is a better predictor than postnatal age for risk of intraventricular hemorrhage in premature babies treated with extracorporeal membrane oxygenation.[17]

2.4 Diseases

Various disorders can arise from defects in the fertilization process.

- Polyspermy results from multiple sperm fertilizing an egg.

However, some researchers have found that in rare pairs of fraternal twins, their origin might have been from the fertilization of one egg cell from the mother and eight sperm cells from the father. This possibility has been investigated by computer simulations of the fertilization process.

2.5 See also

- Spontaneous conception, the unassisted conception of a subsequent child after prior use of assisted reproductive technology

2.6 References

[1] Garrison, Fielding. An Introduction to the History of Medicine, pages 566-567 (Saunders 1921).

[2] http://www.goaskalice.columbia.edu/0116.html

[3] http://www.americanpregnancy.org/preventingpregnancy/pregnancyfaqmyths.html

[4] Lawyers Guide to Forensic Medicine SBN 978-1-85941-159-9 By Bernard Knight - Page 188 "Pregnancy is well known to occur from such external ejaculation ..."

[5] "Fertilization of the Ovum". *Gray's Anatomy*. Retrieved 2010-10-16.

[6] "Fertilization". Retrieved 28 July 2010.

[7] Angier, Natalie (2007-06-12). "Sleek, Fast and Focused: The Cells That Make Dad Dad". *The New York Times*.

[8] Suzanne Wymelenberg, *Science and Babies*, National Academy Press, page 17

[9] Richard E. Jones and Kristin H. Lopez, Human Reproductive Biology, Third Edition, Elsevier, 2006, page 238

[10] "Fertilization: The Cortical Reaction". *Boundless*. Boundless. Retrieved 14 March 2013.

[11] Jodar, M.; Selvaraju, S.; Sendler, E.; Diamond, M. P.; Krawetz, S. A.; for the Reproductive Medicine Network (2013). "The presence, role and clinical use of spermatozoal RNAs". *Human Reproduction Update* **19** (6): 604–624. doi:10.1093/humupd/dmt031. PMC 3796946. PMID 23856356.

[12] Marieb, Elaine M. *Human Anatomy and Physiology, 5th ed.* pp. 1119-1122 (2001). ISBN 0-8053-4989-8

[13] Wagner F, Erdösová B, Kylarová D (December 2004). "Degradation phase of apoptosis during the early stages of human metanephros development". *Biomed Pap Med Fac Univ Palacky Olomouc Czech Repub* **148** (2): 255–6. doi:10.5507/bp.2004.054. PMID 15744391.

[14] Robinson, H. P.; Fleming, J. E. E. (1975). "A Critical Evaluation of Sonar "crown-Rump Length" Measurements". *BJOG: an International Journal of Obstetrics and Gynaecology* **82** (9): 702–710. doi:10.1111/j.1471-0528.1975.tb00710.x.

[15] Geirsson RT (May 1991). "Ultrasound instead of last menstrual period as the basis of gestational age assignment". *Ultrasound Obstet Gynecol* **1** (3): 212–9. doi:10.1046/j.1469-0705.1991.01030212.x. PMID 12797075.

[16] Derived from a standard deviation in this interval of 2.6, as given in: Fehring RJ, Schneider M, Raviele K (2006). "Variability in the phases of the menstrual cycle". *J Obstet Gynecol Neonatal Nurs* **35** (3): 376–84. doi:10.1111/j.1552-6909.2006.00051.x. PMID 16700687.

[17] Alan H. Jobe, MD, PhD. *Post-fertilizational age and IVH in ECMO patients.* RadiologySource Volume 145, Issue 2, Page A2 (August 2004). PII: S0022-3476(04)00583-9. doi:10.1016/j.jpeds.2004.07.010.

2.7   External links

- Fertilization (Conception)

Chapter 3

Oocyte activation

Oocyte (or **ovum/egg**) **activation** is a series of processes that occur in the oocyte during fertilization.

Sperm entry causes calcium release into the oocyte. In mammals, this has been proposed to be caused by the introduction of phospholipase C isoform zeta (PLCζ) from the sperm cytoplasm, although this remains to be established definitively. Activation of the ovum includes the following events:

- Cortical reaction to block against other sperm cells

- Activation of egg metabolism

- Reactivation of meiosis

- DNA synthesis

3.1 Sperm trigger of egg activation

The sperm may trigger egg activation via the interaction between a sperm protein and an egg surface receptor. Izumo is the sperm cell signal, that will trigger the egg receptor Juno.[1] This receptor is activated by the sperm binding and a possible signalling pathway could be the activation of a tyrosine kinase which then activates phospholipase C (PLC). The inositol signaling system has been implicated as the pathway involved with egg activation. **IP$_3$** and **DAG** are produced from the cleavage of **PIP$_2$** by phospholipase C. However, another hypothesis is that a soluble 'sperm factor' diffuses from the sperm into the egg cytosol upon sperm-oocyte fusion. The results of this interaction could activate a signal transduction pathway that uses second messengers. A novel PLC isoform, PLCζ, may be the equivalent of the mammalian sperm factor. A 2002 study demonstrated that mammaliam sperm contain PLC zeta which can start the signaling cascade.[2]

3.2 Fast and slow block to polyspermy

Further information: Cortical reaction

Polyspermy is the condition when multiple sperm fuse with a single egg. This results in duplications of genetic material. In sea urchins, the block to polyspermy comes from two mechanisms: the fast block and the slow block. The fast block is an electrical block to polyspermy. The resting potential of an egg is -70mV. After contact with sperm, an influx of sodium ions increases the potential up to $+20$mV. The slow block is through a biochemical mechanism triggered by a wave of calcium increase. The rise of calcium is both necessary and sufficient to trigger the slow block. In the cortical reaction, cortical granules directly beneath the plasma membrane are released into the space between the plasma membrane and the vitelline membrane (the perivitelline space). An increase in calcium triggers this release. The contents of the granules contain proteases, mucopolysaccharides, hyalin, and peroxidases. The proteases cleave the bridges connecting the plasma membrane and the vitelline membrane and cleave the bindin to release the sperm. The mucopolysaccharides attract water to raise the vitelline membrane. The hyalin forms a layer adjacent to the plasma membrane and the peroxidases cross-link the protein in the vitelline membrane to harden it and make it impenetrable to sperm. Through these molecules the vitelline membrane is transformed into the fertilization membrane or fertilization envelope. In mice, the zona reaction is the equivalent to the cortical reaction in sea urchins. The terminal sugars from ZP3 are cleaved to release the sperm and prevent new binding.

3.3 Reactivation of meiosis

The meiotic cycle of the oocyte was suspended in metaphase of the second meiotic division. Once PLCζ is introduced into the oocyte by the sperm cell, it cleaves phospholipid phosphatidylinositol 4,5-bisphosphate (PIP_2) into diacyl glycerol (DAG) and inositol 1,4,5-trisphosphate (IP_3). In most cells, this occurs at the cell membrane however, evidence suggests that the PIP_2 required for oocyte activation is potentially stored in intracellular vesicles dispersed throughout the cytoplasm.[3] The IP_3 produced then triggers calcium oscillations which reactivate the meiotic cycle. This results in the production and extrusion of the second polar body.[4]

3.4   DNA synthesis

4 hours after fusion of sperm and ovum, DNA synthesis begins.[4] Male and female pronuclei move to the centre of the egg and membranes break down. Male protamines are replaced with histones and the male DNA is demethylated. Chromosomes then orientate on the metaphase spindle for mitosis. This combination of the two genomes is called **syngamy**.[4]

The sperm contributes a pronucleus and a centriole to the egg. Most other components and organelles are rapidly degraded. Mitochondria are rapidly ubiquinated and destroyed. **Oxidative stress theory** is a hypothesis that it is evolutionarily favourable for mitochondria from the father to be destroyed, as it there is a greater possibility that the mitochondrial DNA has become mutated or damaged. This is because mtDNA is not protected by histones and has poor repair mechanisms. Due to the increased metabolic activity of the sperm compared to the egg, due to its motility, there is greater production of reactive oxygen species and therefore greater chance of mutation.[4] Furthermore, sperm are exposed to reactive oxygen species from leukocytes in the epididymis during transit.[4] Additionally, quality control of spermatozoa is much worse than for the ovum, as many sperm are released whereas only one dominant follicle is released per cycle. This competitive selection helps to ensure the most 'fit' ova are selected for fertilisation.[4]

3.5   Artificial oocyte activation

Oocyte activation may be artificially facilitated by calcium ionophores, something that is speculated to be useful in case of fertilization failure, such as still occurs in 1–5% of intracytoplasmic sperm injection (ICSI) cycles.[5] Another of method is by using the drug Roscovitine, this reduces the activity of M-phase promoting factor activity in mice.[6]

3.6   References

[1] Bianchi E, Doe B , Goulding D , Wright GJ (2014). "Juno is the egg Izumo receptor and is essential for mammalian fertilization". *Nature* **508** (7497): 483. doi:10.1038/nature13203.

[2] Saunders C, Larman M, Parrington J, Cox L, Royse J, Blayney L, Swann K, Lai F (2002). "PLC zeta: a sperm-specific trigger of Ca(2+) oscillations in eggs and embryo development.". *Development* **129** (15): 3533–44. PMID 12117804.

[3] Yu, Yuansong; Nomikos, Michail; Theodoridou, Maria; Nounesis, George; Lai, F. Anthony; Swann, Karl (2012-01-15). "PLCζ causes Ca2+ oscillations in mouse eggs by targeting intracellular and not plasma membrane PI(4,5)P2". *Molecular Biology of the Cell* **23** (2): 371–380. doi:10.1091/mbc.E11-08-0687. ISSN 1059-1524. PMC 3258180. PMID 22114355.

[4] Johnson, M. (2007). *Essential Reproduction* (6th ed.). Oxford: Blackwell. ISBN 9781405118668.

[5] Kashir, J.; Heindryckx, B.; Jones, C.; De Sutter, P.; Parrington, J.; Coward, K. (2010). "Oocyte activation, phospholipase C zeta and human infertility". *Human Reproduction Update* **16** (6): 690–703. doi:10.1093/humupd/dmq018. PMID 20573804.

[6] Iba T, Yano Y, Umeno M, Hinokio K, Kuwahara A, Irahara M, Yamano S and Yasui T. (2011) Roscovitine in combination with calcium ionophore induces oocyte activation through reduction of M-phase promoting factor activity in mice. *Zygote* **20**:321-325. PMID 22008472

Chapter 4

Zygote

For other uses, see Zygote (disambiguation).
"Fertilized egg" redirects here. For the food product, see Balut (egg).

A **zygote** (from Greek ζυγωτός *zygōtos* "joined" or "yoked", from ζυγοῦν *zygoun* "to join" or "to yoke"),[1] is a eukaryotic cell formed by a fertilization event between two gametes. The zygote's genome is a combination of the DNA in each gamete, and contains all of the genetic information necessary to form a new individual. In multicellular organisms, the zygote is the earliest developmental stage. In single-celled organisms, the zygote can divide asexually by mitosis to produce identical offspring.

Oscar Hertwig and Richard Hertwig made some of the first discoveries on animal zygote formation.

4.1   Fungi

In fungi, the sexual fusion of haploid cells is called karyogamy. The result of karyogamy is a diploid cell called a zygote or zygospore. This cell may then enter meiosis or mitosis depending on the life cycle of the species.

4.2   Plants

In plants, the zygote may be polyploid if fertilization occurs between meiotically unreduced gametes.

In land plants, the zygote is formed within a chamber called the archegonium. In seedless plants, the archegonium is usually flask-shaped, with a long hollow neck through which the sperm cell enters. As the zygote divides and grows, it does so inside the archegonium.

4.3   Humans

In human fertilization, two 1n haploid cells—an ovum (female gamete) and a sperm cell (male gamete)—combine to form a single 2n diploid cell called the zygote. DNA is then replicated in the two separate pronuclei derived from the sperm and ovum, making the zygote's chromosome number temporarily 4n diploid. After approximately 30 hours, fusion of the pronuclei and subsequent mitotic division produce two 2n diploid daughter cells called blastomeres.[2]

Between the stages of fertilization and implantation, the developing human is called the *preimplantation conceptus* or the proembryo. It is not correct to call the conceptus an *embryo*, because it will later differentiate into both intraembryonic and extraembryonic tissues,[3] and can even split to produce multiple embryos (identical twins).

After fertilization, the conceptus travels down the oviduct towards the uterus while continuing to divide[4] mitotically without actually increasing in size, in a process called cleavage.[5] After four divisions, the conceptus consists of 16 blastomeres, and it is known as the morula.[6] Through the processes of compaction, cell division, and blastulation, the conceptus takes the form of the blastocyst by the fifth day of development, just as it approaches the site of implantation.[7] When the blastocyst hatches from the zona pellucida, it can implant in the endometrial lining of the uterus and begin the embryonic stage of development.

The human zygote has been genetically edited in experiments designed to cure inherited diseases.[8]

4.4   In other species

A Chlamydomonas zygote that contains chloroplast DNA (cpDNA) from both parents, such cells generally are rare since normally cpDNA is inherited uniparental from the mt+ mating type parent. These rare biparental zygotes allowed mapping of chloroplast genes by recombination.

4.5   In Protozoa

In the Amoeba, reproduction occurs by cell division of the parent cell: first the nucleus of the parent divides into two and then the cell membrane also cleaves, becoming two "daughter" Amoebae.

4.6   See also

- Proembryo

4.7   References

[1] "English etymology of zygote". *myetymology.com.*

[2] Blastomere Encyclopædia Britannica. Encyclopædia Britannica Online. Encyclopædia Britannica Inc., 2012. Web. 06 Feb. 2012.

[3] Larsen's Human Embryology. 4th Ed. Page 4.

[4] O'Reilly, Deirdre. "Fetal development". *MedlinePlus Medical Encyclopedia* (2007-10-19). Retrieved 2009-02-15.

[5] Klossner, N. Jayne and Hatfield, Nancy. *Introductory Maternity & Pediatric Nursing,* p. 107 (Lippincott Williams & Wilkins, 2006).

[6] Neas, John F. "Human Development". *Embryology Atlas*

[7] Blackburn, Susan. *Maternal, Fetal, & Neonatal Physiology,* p. 80 (Elsevier Health Sciences 2007).

[8] Human zygote edited genetically

Chapter 5

Cleavage (embryo)

In embryology, **cleavage** is the division of cells in the early embryo. The zygotes of many species undergo rapid cell cycles with no significant growth, producing a cluster of cells the same size as the original zygote. The different cells derived from cleavage are called blastomeres and form a compact mass called the morula. Cleavage ends with the formation of the blastula.

Depending mostly on the amount of yolk in the egg, the cleavage can be **holoblastic** (total or entire cleavage) or **meroblastic** (partial cleavage). The pole of the egg with the highest concentration of yolk is referred to as the vegetal pole while the opposite is referred to as the animal pole.

Cleavage differs from other forms of cell division in that it increases the number of cells without increasing the mass. This means that with each successive subdivision, the ratio of nuclear to cytoplasmic material increases.[1]

5.1 Mechanism

The rapid cell cycles are facilitated by maintaining high levels of proteins that control cell cycle progression such as the cyclins and their associated cyclin-dependent kinases (cdk). The complex Cyclin B/cdc2 a.k.a. MPF (maturation promoting factor) promotes entry into mitosis.

The processes of karyokinesis (mitosis) and cytokinesis work together to result in cleavage. The mitotic apparatus is made up of a central spindle and polar asters made up of polymers of tubulin protein called microtubules. The asters are nucleated by centrosomes and the centrosomes are organized by centrioles brought into the egg by the sperm as basal bodies. Cytokinesis is mediated by the contractile ring made up of polymers of actin protein called microfilaments. Karyokinesis and cytokinesis are independent but spatially and temporally coordinated processes. While mitosis can occur in the absence of cytokinesis, cytokinesis requires the mitotic apparatus.

The end of cleavage coincides with the beginning of zygotic transcription. This point is referred to as the midblastula transition and appears to be controlled by the nuclear: cytoplasmic ratio (about 1/6).

5.2 Types of cleavage

5.2.1 Determinate

Determinate cleavage (also called mosaic cleavage) is in most protostomes. It results in the developmental fate of the cells being set early in the embryo development. Each blastomere produced by early embryonic cleavage does not have the capacity to develop into a complete embryo.

5.2.2 Indeterminate

A cell can only be indeterminate (also called regulative) if it has a complete set of undisturbed animal/vegetal cytoarchitectural features. It is characteristic of deuterostomes – when the original cell in a deuterostome embryo divides, the two resulting cells can be separated, and each one can individually develop into a whole organism.

5.2.3 Holoblastic

In the absence of a large concentration of yolk, four major cleavage types can be observed in isolecithal cells (cells with a small even distribution of yolk) or in mesolecithal cells (moderate amount of yolk in a gradient) – **bilateral** holoblastic, **radial** holoblastic, **rotational** holoblastic, and **spiral** holoblastic, cleavage.[2] These holoblastic cleavage planes pass all the way through isolecithal zygotes during the process of cytokinesis. Coeloblastula is the next stage of development for eggs that undergo these radial cleaving. In holoblastic eggs, the first cleavage always occurs along the vegetal-animal axis of the egg, the second cleavage is perpendicular to the first. From here, the spatial arrangement of blastomeres can follow various patterns, due to different planes of cleavage, in various organisms.

- Bilateral

  The first cleavage results in bisection of the zygote into left and right halves. The following cleavage planes are centered on this axis and result in the two halves being mirror images of one another. In bilateral holoblastic cleavage, the divisions of the blastomeres are complete and separate; compared with bilateral meroblastic cleavage, in which the blastomeres stay partially connected.

- Radial

  Radial cleavage is characteristic of the deuterostomes, which include some vertebrates and echinoderms, in which the spindle axes are parallel or at right angles to the polar axis of the oocyte.

- Rotational

  Mammals display rotational cleavage, and an isolecithal distribution of yolk (sparsely and evenly distributed). Because the cells have only a small amount of yolk, they require immediate implantation onto the uterine wall in order to receive nutrients.

  Rotational cleavage involves a normal first division along the meridional axis, giving rise to two daughter cells. The way in which this cleavage differs is that one of the daughter cells divides meridionally, whilst the other divides equatorially.

- Spiral

  Spiral cleavage is conserved between many members of the lophotrochozoan taxa, referred to as Spiralia.[3] Most spiralians undergo equal spiral cleavage, although some undergo unequal cleavage (see below).[4] This group includes annelids, molluscs, and sipuncula. Spiral cleavage can vary between species, but generally the first two cell divisions result in four macromeres, also called blastomeres, (A, B, C, D) each representing one quadrant of the embryo. These first two cleavages are oriented in planes that occur

at right angles parallel to the animal-vegetal axis of the zygote.[3] At the 4-cell stage, the A and C macromeres meet at the animal pole, creating the animal cross-furrow, while the B and D macromeres meet at the vegetal pole, creating the vegetal cross-furrow.[5] With each successive cleavage cycle, the macromeres give rise to quartets of smaller micromeres at the animal pole.[6][7] The divisions that produce these quartets occur at an oblique angle, an angle that is not a multiple of 90°, to the animal-vegetal axis.[7] Each quartet of micromeres is rotated relative to their parent macromere, and the chirality of this rotation differs between odd and even numbered quartets, meaning that there is alternating symmetry between the odd and even quartets.[3] In other words, the orientation of divisions that produces each quartet alternates between being clockwise and counterclockwise with respect to the animal pole.[7] The alternating cleavage pattern that occurs as the quartets are generated produces quartets of micromeres that reside in the cleavage furrows of the four macromeres.[5] When viewed from the animal pole, this arrangement of cells displays a spiral pattern.

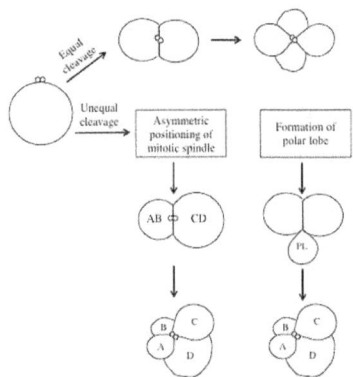

D quadrant specification through equal and unequal cleavage mechanisms. At the 4-cell stage of equal cleavage, the D macromere has not been specified yet. It will be specified after the formation of the third quartet of micromeres. Unequal cleavage occurs in two ways: asymmetric positioning of the mitotic spindle, or through the formation of a polar lobe (PL).

Specification of the D macromere and is an important aspect of spiralian development. Although the primary axis, animal-vegetal, is determined during oogenesis, the secondary axis, dorsal-ventral, is determined by the specification of the D quadrant.[7] The D macromere facilitates cell divisions that differ from those produced by the other three macromeres. Cells of the D quadrant give rise to dorsal and posterior structures of the spiralian.[7] Two known mechanisms exist to specify the D quadrant. These mechanisms include equal cleavage and unequal cleavage.

In equal cleavage, the first two cell divisions produce four macromeres that are indistinguishable from one another. Each macromere has the potential of becoming the D macromere.[6] After the formation of the third quartet, one of the macromeres initiates maximum contact with the overlying micromeres in the animal pole of the embryo.[6][7] This contact is required to distinguish one macromere as the official D quadrant blastomere. In equally cleaving spiral embryos, the D quadrant is not specified until after the formation of the third quartet, when contact with the micromeres dictates one cell to become the future D blastomere. Once specified, the D blastomere signals to surrounding micromeres to lay out their cell fates.[7]

In unequal cleavage, the first two cell divisions are unequal producing four cells in which one cell is bigger than the other three. This larger cell is specified as the D macromere.[6][7] Unlike equally cleaving spiralians, the D macromere is specified at the four-cell stage during unequal cleavage. Unequal cleavage can occur in two ways. One method involves asymmetric positioning of the cleavage spindle.[7] This occurs when the aster at one pole attaches to the cell membrane, causing it to be much smaller than the aster at the other pole.[6] This results in an unequal cytokinesis, in which both macromeres inherit part of the animal region of the egg, but only the bigger macromere inherits the vegetal region.[6] The second mechanism of unequal cleavage involves the production of an enucleate, membrane bound, cytoplasmic protrusion, called a polar lobe.[6] This polar lobe forms at the vegetal pole during cleavage, and then gets shunted to the D blastomere.[5][6] The polar lobe contains vegetal cytoplasm, which becomes inherited by the future D macromere.[7]

5.2.4 Meroblastic

In the presence of a large amount of yolk in the fertilized egg cell, the cell can undergo partial, or meroblastic, cleav-

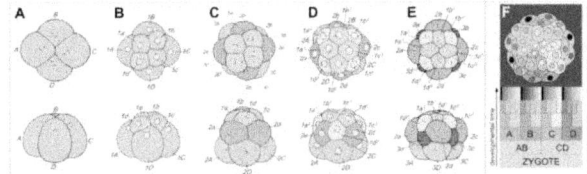

Spiral cleavage in marine snail of the genus Trochus.

age. Two major types of meroblastic cleavage are **discoidal** and **superficial**.[8]

- Discoidal

  In discoidal cleavage, the cleavage furrows do not penetrate the yolk. The embryo forms a disc of cells, called a blastodisc, on top of the yolk. Discoidal cleavage is commonly found in monotremes, birds, reptiles, and fish that have telolecithal egg cells (egg cells with the yolk concentrated at one end).

- Superficial

  In superficial cleavage, mitosis occurs but not cytokinesis, resulting in a polynuclear cell. With the yolk positioned in the center of the egg cell, the nuclei migrate to the periphery of the egg, and the plasma membrane grows inward, partitioning the nuclei into individual cells. Superficial cleavage occurs in arthropods that have centrolecithal egg cells (egg cells with the yolk located in the center of the cell).

5.3 Placentals

Differences exist between the cleavage in placental mammals and the cleavage in other animals. Mammals have a slow rate of division that is between 12 and 24 hours. These cellular divisions are asynchronous. Zygotic transcription starts at the two-, four-, or eight-cell stage. Cleavage is holoblastic and rotational.

At the eight-cell stage, the embryo goes through some changes. Most of the blastomeres in this stage become polarized and develop tight junctions with the other blastomeres. This process leads to the development of two different populations of cells: Polar cells on the outside and apolar cells on the inside. The outer cells, called the trophoblast cells, pump sodium in from the outside, which automatically brings water in with it to the basal (inner) surface to form a blastocoel cavity in a process called com-

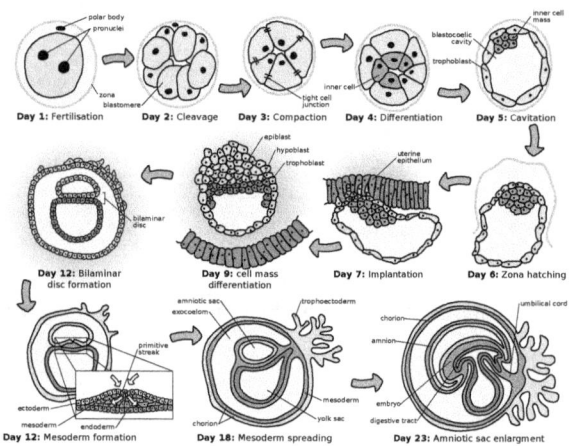

The initial stages of human embryogenesis.

paction. The embryo is now called a blastocyst. The trophoblast cells will eventually give rise to the embryonic contribution to the placenta called the chorion. The inner cells are pushed to one side of the cavity (because the embryo isn't getting any bigger) to form the inner cell mass (ICM) and will give rise to the embryo and some extraembryonic membranes. At this stage, the embryo is called a blastocyst.

5.4 See also

- Embryogenesis

- Blastocyst

5.5 References

[1] Forgács, G. & Newman, Stuart A. (2005). "Cleavage and blastula formation". *Biological physics of the developing embryo*. Cambridge University Press. p. 27. ISBN 978-0-521-78337-8.

[2] "Early Development of the Nematode Caenorhabditis elegans". Retrieved 2007-09-17.

[3] Shankland, M.; Seaver, E. C. (2000). "Evolution of the bilaterian body plan: What have we learned from annelids?". *Proceedings of the National Academy of Sciences* **97** (9): 4434–7. Bibcode:2000PNAS...97.4434S. doi:10.1073/pnas.97.9.4434. JSTOR 122407. PMC 34316. PMID 10781038.

[4] Henry, J. (2002). "Conserved Mechanism of Dorsoventral Axis Determination in Equal-Cleaving Spiralians". *Developmental Biology* **248** (2): 343–355. doi:10.1006/dbio.2002.0741. PMID 12167409.

[5] Boyer, Barbara C.; Jonathan, Q. Henry (1998). "Evolutionary Modifications of the Spiralian Developmental Program". *Integrative and Comparative Biology* **38** (4): 621–33. doi:10.1093/icb/38.4.621. JSTOR 4620189.

[6] Freeman, Gary; Lundelius, Judith W. (1992). "Evolutionary implications of the mode of D quadrant specification in coelomates with spiral cleavage". *Journal of Evolutionary Biology* **5** (2): 205–47. doi:10.1046/j.1420-9101.1992.5020205.x.

[7] Lambert, J.David; Nagy, Lisa M (2003). "The MAPK cascade in equally cleaving spiralian embryos". *Developmental Biology* **263** (2): 231–41. doi:10.1016/j.ydbio.2003.07.006. PMID 14597198.

[8] "Current Notes". Retrieved 2007-09-17.

[9] Gilbert SF (2003). *Developmental biology* (7th ed.). Sinauer. p. 214. ISBN 0-87893-258-5.

[10] Kardong, Kenneth V. (2006). *Vertebrates: Comparative Anatomy, Function, Evolution* (4th ed.). McGraw-Hill. pp. 158–64.

5.6 Bibliography

- Wilt, F. & Hake, S. (2004). *Principles of Developmental Biology*.

- Scott F. Gilbert (2003). *Developmental Biology*.

5.7 External links

- Valentine, James W. (1997). "Cleavage Patterns and the Topology of the Metazoan Tree of Life". *Proceedings of the National Academy of Sciences of the United States of America* **94** (15): 8001–5. Bibcode:1997PNAS...94.8001V. doi:10.1073/pnas.94.15.8001. PMC 21545. PMID 9223303.

- 'What are the 'advantages' of developing a deuterostome pattern of embryonic' on MadSci Network

- Lee, Seung-Cheol; Mietchen, Daniel; Cho, Jee-Hyun; Kim, Young-Sook; Kim, Cheolsu; Hong, Kwan Soo; Lee, Chulhyun; Kang, Dongmin; Lee, Wontae; Cheong, Chaejoon (200 7). "In vivo magnetic resonance microscopy of differentiation in Xenopus laevis embryos from the first cleavage onwards". *Differentiation* **75**. doi:10.1111/j.1432-0436.2006.00114.x. Check date values in: |date= (help)

Chapter 6

Blastomere

In biology, a **blastomere** is a type of cell produced by cleavage (cell division) of the zygote after fertilization and is an essential part of blastula formation.[1]

6.1   See also

- Blastocoel

- Blastocyst

- Oocyte

6.2   References

[1] Blastomere Encyclopædia Britannica. Encyclopædia Britannica Online. Encyclopædia Britannica Inc., 2012. Web. 06 Feb. 2012.

- "Blastomere." *Stedman's Medical Dictionary, 27th ed.* (2000). ISBN 0-683-40007-X

- Moore, Keith L. and T.V.N. Persaud. *The Developing Human: Clinically Oriented Embryology, 7th Ed.* (2003). ISBN 0-7216-9412-8

Chapter 7

Morula

For the South African football player, see Lebohang Morula.

A **morula** (Latin, *morus*: mulberry) is an early stage embryo consisting of cells (called blastomeres) in a solid ball contained within the zona pellucida.[1][2]

A morula is distinct from a blastocyst in that a morula (3-4 days post fertilization) is an 16 cell mass in a spherical shape whereas a blastocyst (4-5 days post fertilization) has a cavity inside the zona pellucida along with an inner cell mass. A morula, if untouched and allowed to remain implanted, will eventually develop into a blastocyst.[3]

The morula is produced by a series of cleavage divisions of the early embryo, starting with the single-celled zygote. Once the embryo has divided into 16 cells, it begins to resemble a mulberry, hence the name *morula* (Latin, *morus*: mulberry).[4] Within a few days after fertilization, cells on the outer part of the morula become bound tightly together with the formation of desmosomes and gap junctions, becoming nearly indistinguishable. This process is known as compaction.[5][6] A cavity forms inside the morula, by the active transport of sodium ions from trophoblast cells and osmosis of water. This results in a hollow ball of cells known as the blastocyst.[7][8] The blastocyst's outer cells will become the first embryonic epithelium (the trophectoderm). Some cells, however, will remain trapped in the interior and will become the inner cell mass (ICM), and are pluripotent. In mammals (except monotremes), the ICM will ultimately form the "embryo proper", while the trophectoderm will form the placenta and extra-embryonic tissues. However, reptiles have a different ICM. The stages are prolonged and divided in 4 parts.[9][10][11][12]

7.1 See also

- Cleavage (embryo)

- Blastula

7.2 References

[1] Boklage, Charles E. (2009). *How New Humans Are Made: Cells and Embryos, Twins and Chimeras, Left and Right, Mind/Self/Soul, Sex, and Schizophrenia.* World Scientific. p. 217. ISBN 9789812835130.

[2] "The Early Embryology of the Chick". UNSW Embryology. Retrieved 2015-03-03.

[3] "The Morula and Blastocyst". the Endowment for Human Development. Retrieved 11 April 2015.

[4] Sherman, Lawrence S. et al., eds. (2001). *Human embryology* (3rd ed.). Elsevier Health Sciences. p. 20. ISBN 978-0-443-06583-5.

[5] Chard, Tim & Lilford, Richard (1995). *Basic sciences for obstetrics and gynaecology.* Springer. p. 18. ISBN 978-3-540-19903-8.

[6] Mercader, Amparo et al. (2008). "Human embryo culture". In Lanza, Robert & Klimanskaya, Irina. *Essential stem cell methods.* Academic Press. p. 343. ISBN 978-0-12-374741-9.

[7] Patestas, Maria Antoniou & Gartner, Leslie P. (2006). *A textbook of neuroanatomy.* Wiley-Blackwell. p. 11. ISBN 978-1-4051-0340-4.

[8] Geisert, R.D. & Malayer, J.R. (2000). "Implantation: Blastocyst formation". In Hafez, B. & Hafez, Elsayed S.E. *Reproduction in farm animals.* Wiley-Blackwell. p. 118. ISBN 978-0-683-30577-7.

[9] Morali, Olivier G. et al. (2005). "Epithelium-Mesenchyme Transitions are Crucial Morphogenetic Events Occurring During Early Development". In Savagner, Pierre. *Rise and fall of epithelial phenotype: concepts of epithelial-mesenchymal transition.* Springer. p. 16. ISBN 978-0-306-48239-7.

[10] Birchmeier, Carmen; et al. (1997). "Morphogenesis of epithelial cells". In Paul, Leendert C. & Issekutz, Thomas B. *Adhesion molecules in health and disease.* CRC Press. p. 208. ISBN 978-0-8247-9824-6.

[11] Nagy, András (2003). *Manipulating the mouse embryo: a laboratory manual*. CSHL Press. pp. 60–61. ISBN 978-0-87969-591-0.

[12] Connell, R.J. & Cutner, A. (2001). "Basic Embryology". In Cardozo, Linda & Staskin, David. *Textbook of female urology and urogynaecology*. Taylor & Francis. p. 92. ISBN 978-1-901865-05-9.

7.3   Further reading

- "Regulative development in mammals"

Chapter 8

Blastocoel

A **blastocoel** (alt. spelling *blastocoele, blastocele*),[1] also termed the **blastocyst cavity**[2] (or *cleavage* or *segmentation cavity*[1]) is the name given to the fluid-filled cavity of the blastula (blastocyst) that results from cleavage of the oocyte (ovum) after fertilization.[1][2] It forms during embryogenesis, as what has been termed a "Third Stage" after the single-celled fertilized oocyte (zygote, ovum) has divided into 16-32 cells,[2] via the process of mitosis. It can be described as the first cell cavity formed as the embryo enlarges, the essential precursor for the differentiated, topologically distinct, gastrula. The adjectival form of blastocoel is *blastocoelic*.[1][3][4][5]

8.1 References

[1] Dorlands Staff (2004). "blastocoele [dictionary entry]". *Dorland's Illustrated Medical Dictionary* (online). Amsterdam, NDE: Elsevier-Saunders. Retrieved 30 January 2016. blastocoele... [blasto- + -coele] the fluid-filled cavity of the mass of cells (blastula) produced by cleavage of a fertilized ovum. Sometimes spelled... [c]alled... [Also] blastocoelic... pertaining to the blastocoele.

[2] Senn, A.; Schöni-Affolter, F.; Dubuis-Grieder, C. & Strauch, E.; et al. (2007). "Module 8, Embryonic Phase; [Section] 8.1 The Carnegie Stages; Synoptic Table of the Carnegie Stages 1 - 6; [page] 'Stage 3, Approx. 4th - 5th day, 0.1 - 0.2 mm' [rev. 25.04.07]". In Manuèle Adé-Damilano. *embryology.ch: Online Course in Embryology for Medicine Students*. Fribourg, CHE: Swiss Virtual Campus, l'Université de Fribourg, et al. Retrieved 30 January 2016. Drs Senn and Dubuis-Grieder are at l'Université Lausanne, Drs Adé-Damilano and Schöni-Affolter at l'Université de Fribourg, and Dr Strauch at Universität Bern.

[3] Ereskovsky, Alexander V. (2010). *The Comparative Embryology of Sponges*. Springer. ISBN 978-90-481-8574-0.

[4] Mader S. S. (2000): Human biology. McGraw-Hill, New York, ISBN 0-07-290584-0; ISBN 0-07-117940-2.

[5] Gilbert S. F. (2010). *Developmental Biology* (Ninth ed.). Sinauer Associates. ISBN 978-0-87893-558-1.

8.2 See also

- Embryo
- Blastula

Chapter 9

Blastocyst

For the non-species specific developmental stage, see Blastula. For the single-celled parasite, see Blastocystis.

The **blastocyst** is a structure formed in the early development of mammals. It possesses an inner cell mass (ICM) which subsequently forms the embryo. The outer layer of the blastocyst consists of cells collectively called the trophoblast. This layer surrounds the inner cell mass and a fluid-filled cavity known as the blastocoele. The trophoblast gives rise to the placenta. The name "blastocyst" arises from the Greek βλαστός *blastos* ("a sprout") and κύστις *kystis* ("bladder, capsule").

In humans, blastocyst formation begins about 5 days after fertilization, when a fluid-filled cavity opens up in the morula, a ball consisting of a few dozen cells. The blastocyst has a diameter of about 0.1-0.2 mm and comprises 200-300 cells following rapid cleavage (cell division). After about 1 day (5–6 days post-fertilization), which is the time usually required to reach the uterus, the blastocyst begins to embed itself into the endometrium of the uterine wall where it will undergo later developmental processes, including gastrulation. Embedding of the blastocyst into the endometrium requires that it hatches from the zona pellucida, which prevents it from adhering to the oviduct as it makes its way to the uterus. The blastocyst is completely embedded in the endometrium only 11–12 days after fertilization.

The use of blastocysts in in-vitro fertilization (IVF) involves culturing a fertilized egg for five days before implanting it into the uterus. It can be a more viable method of fertility treatment than traditional IVF. The inner cell mass of blastocysts is also a source of embryonic stem cells.

9.1 Development cycle

During human embryogenesis, the blastocyst arises from the morula in the uterus, 5 days after fertilization. The early embryo undergoes cell differentiation and structural changes to become the blastocyst. It is then prepared for implantation into the uterine wall 6 days after fertilization. Implantation marks the end of the germinal stage of embryogenesis.[1]

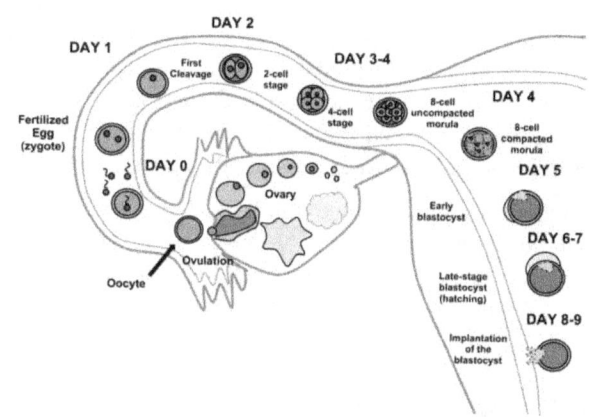

Early development of the embryo from ovulation through implantation in humans. The blastocyst stage occurs between 5 and 8-9 days following conception.

9.1.1 Blastocyst formation

The morula, which precedes the blastocyst, is an early embryo composed of 16 undifferentiated cells. Shortly following the morula's entry into the uterus from the Fallopian tube, the morula becomes the blastocyst through cellular differentiation and cavitation. The morula's cells differentiate into two types: an inner cell mass growing on the interior of the blastocoel and trophoblast cells growing on the exterior.[2] The animal pole refers to the side of the blastocyst where the ICM resides, while the vegetal pole is on the opposite side. Cavitation is the process by which a fluid cavity forms inside the embryo. The trophoblast cells pump sodium ions into the center of the embryo, which causes water to enter through osmosis. This forms an internal fluid-filled cavity called the blastocoel. This distinguishable blastocoel cavity in addition to cellular specification are both hallmark identities of the blastocyst.[3]

9.1.2 Implantation

Implantation is critical to the survival and development of the early embryo. It establishes a connection between the mother and the early embryo which will continue through the remainder of the pregnancy. Implantation is made possible through structural changes in both the blastocyst and endometrial wall.[4] The zona pellucida surrounding the blastocyst breaches, referred to as hatching. This removes the constraint on the physical size of the embryonic mass and exposes the outer cells of the blastocyst to the interior of the uterus. Furthermore, hormonal changes in the mother, specifically a peak in luteinizing hormone (LH) prepares the endometrium to receive the blastocyst and envelope it. Once bound to the extracellular matrix of the endometrium, trophoblast cells secrete enzymes and other factors to embed the blastocyst into the uterine wall. The enzymes released degrade the endometrial lining, while autocrine growth factors such as human chorionic gonadotropin (hCG) and insulin-like growth factor (IGF) allow the blastocyst to further invade the endometrium.[5]

Implantation in the uterine wall allows for the next step in embryogenesis, gastrulation, which includes formation of the placenta from trophoblastic cells and differentiation of the ICM into the amniotic sac and epiblast.

9.2 Structure

The blastocyst is made up of cells from the inner cell mass and the blastocoel.

There are two types of blastomere cells:[6]

- The inner cell mass, also known as the embryoblast, gives rise to the primitive endoderm and the epiblast.

  - The primitive endoderm develops into the amniotic sac which forms the fluid-filled cavity that the embryo resides in during pregnancy.[7]
  - The epiblast gives rise to the three germ layers of the developing embryo during gastrulation (endoderm, mesoderm, and ectoderm).

- The trophoblast is a layer of cells forming the outer ring of the blastocyst that combines with the maternal endometrium to form the placenta. Trophoblast cells also secrete factors to make the blastocoel.[8]

  - Cytotrophoblast is the inner layer of the trophoblast, composed of stem cells which give rise

to cells comprising the chorionic villi, placenta, and syncytiotrophoblast.

- Syncytiotrophoblast is the outermost layer of the trophoblast. These cells secrete proteolytic enzymes to break down the endometrial extracellular matrix to allow for implantation of the blastocyst in the uterine wall.[9]

The blastocoel fluid cavity contains amino acids, growth factors, and other necessary molecules for cellular differentiation.[10]

9.2.1 Cell specification

Multiple processes control cell lineage specification in the blastocyst to produce the trophoblast, epiblast, and primitive endoderm. These processes include: gene expression, cell signaling, cell-cell contact and positional relationships, and epigenetics.

Once the ICM has been established within the blastocyst, this cell mass prepares for further specification into the epiblast and primitive endoderm. This process of specification is determined in part by fibroblast growth factor (FGF) signaling which generates a MAP kinase pathway to alter cellular genomes.[11] Further segregation of blastomeres into the trophoblast and inner cell mass are regulated by the homeodomain protein, Cdx2. This transcription factor represses the expression of Oct4 and Nanog transcription factors in the trophectoderm.[12] These genomic alterations allow for the progressive specification of both epiblast and primitive endoderm lineages at the end of the blastocyst phase of development preceding gastrulation.

Trophoblasts express integrin on their cell surfaces which allow for adhesion to the extracellular matrix of the uterine wall. This interaction allows for implantation and also triggers further specification into the three different cell types, preparing the blastocyst for gastrulation.[13]

9.3 Clinical implications

9.3.1 Pregnancy tests

Levels of human chorionic gonadotropin secreted by the blastocyst during implantation is the factor measured in a pregnancy test. HCG can be measured in both the blood and urine to determine if a woman is pregnant. More hCG is secreted in a multiple pregnancy. Blood tests of hCG can also be used to check for abnormal pregnancies.[14]

9.3.2 *In vitro* fertilization

In vitro fertilization is an alternative to traditional *in vivo* fertilization for fertilizing an egg with sperm and implanting that embryo into a female's womb. For many years the embryo was inserted into the fallopian tube two to three days after fertilization. However at this stage of development it is very difficult to predict which embryos will develop best, and several embryos were typically implanted. Several implanted embryos helped to guarantee that there would be a developing fetus but it also led to the development of multiple fetuses. This was a major problem and drawback for using embryos to IVF.

A recent breakthrough for in vitro fertilization is the use of blastocysts. A blastocyst would be implanted five to six days after the eggs had been fertilized.[15] After five or six days it is much easier to determine which embryos will result in healthy live births. Knowing which embryos will succeed allows just one blastocyst to be implanted, cutting down dramatically on the health risk and expense of multiple births. Now that the nutrient sources for embryonic and blastocyst development has been determined, it is much easier to give embryos the correct nutrients in order to sustain them into the blastocyst phase. Blastocyst implantation through in vitro fertilization is a painless procedure in which a catheter is inserted into the vagina, guided through the cervix via ultrasound, into the uterus where the blastocysts are inserted into the womb.

Blastocysts also offer an advantage because they can be used to genetically test the cells to check for genomic problems. There are enough cells in a blastocyst that a few trophectoderm cells are able to be removed without disturbing the developing blastocyst. These cells can be tested for chromosome aneuploidy using preimplantation genetic screening (PGS).

9.4 See also

- Developmental biology

- Human embryogenesis

9.5 References

This article incorporates text in the public domain from the 20th edition of Gray's Anatomy (1918)

[1] Sherk, Stephanie Dionne (2006). "Prenatal Development". *Gale Encyclopedia of Children's Health*. Retrieved 2013-12-07.

[2] Clinic, Mayo (2012). "Fetal development: The first trimester". *Mayo Foundation for Medical Education*. Retrieved 2013-12-07.

[3] Gilbert SF. Developmental Biology. 6th edition. Sunderland (MA): Sinauer Associates; 2000. Early Mammalian Development. Available from: http://www.ncbi.nlm.nih.gov/books/NBK10052/

[4] Zhang, Shuang; Lin, Haiyan; Kong, Shuangbo; Wang, Shumin; Wang, Hongmei; Wang, Haibin; Armant, D. Randall (2013). "Physiological and molecular determinants of embryo implantation". *Molecular Aspects of Medicine* **34** (5): 939–80. doi:10.1016/j.mam.2012.12.011. PMID 23290997.

[5] Srisuparp, Santha; Strakova, Zuzana; Fazleabas, Asgerally T (2001). "The Role of Chorionic Gonadotropin (CG) in Blastocyst Implantation". *Archives of Medical Research* **32** (6): 627–34. doi:10.1016/S0188-4409(01)00330-7. PMID 11750740.

[6] Scott F. Gilbert (15 July 2013). *Developmental Biology*. Sinauer Associates, Incorporated. ISBN 978-1-60535-173-5.

[7] Schoenwolf, Gary C., and William J. Larsen. *Larsen's Human Embryology*. 4th ed. Philadelphia: Churchill Livingstone/Elsevier, 2009. Print.

[8] James, J. L; Stone, PR; Chamley, LW (2005). "Cytotrophoblast differentiation in the first trimester of pregnancy: Evidence for separate progenitors of extravillous trophoblasts and syncytiotrophoblast". *Reproduction* **130** (1): 95–103. doi:10.1530/rep.1.00723. PMID 15985635.

[9] Vićovac, L; Aplin, JD (1996). "Epithelial-mesenchymal transition during trophoblast differentiation". *Acta anatomica* **156** (3): 202–16. doi:10.1159/000147847. PMID 9124037.

[10] Gasperowicz, M.; Natale, D. R. C. (2010). "Establishing Three Blastocyst Lineages--Then What?". *Biology of Reproduction* **84** (4): 621–30. doi:10.1095/biolreprod.110.085209. PMID 21123814.

[11] Yamanaka, Y.; Lanner, F.; Rossant, J. (2010). "FGF signal-dependent segregation of primitive endoderm and epiblast in the mouse blastocyst". *Development* **137** (5): 715–24. doi:10.1242/dev.043471. PMID 20147376.

[12] Strumpf, D.; Mao, CA; Yamanaka, Y; Ralston, A; Chawengsaksophak, K; Beck, F; Rossant, J (2005). "Cdx2 is required for correct cell fate specification and differentiation of trophectoderm in the mouse blastocyst". *Development* **132** (9): 2093–102. doi:10.1242/dev.01801. PMID 15788452.

[13] C.H. Damsky; Librach, C; Lim, KH; Fitzgerald, ML; McMaster, MT; Janatpour, M; Zhou, Y; Logan, SK; Fisher, SJ (1994-12-01). "Integrin switching regulates normal trophoblast invasion". *Development* **120** (12): 3657–66. PMID 7529679.

[14] "Human Chorionic Gonadotropin (hCG)". *WebMD*. 2010. Retrieved 2013-12-07.

[15] Fong, C. Y.; Bongso, A.; Ng, S. C.; Anandakumar, C.; Trounson, A.; Ratnam, S. (1997). "Ongoing normal pregnancy after transfer of zona-free blastocysts: Implications for embryo transfer in the human". *Human Reproduction* **12** (3): 557–60. doi:10.1093/humrep/12.3.557. PMID 9130759.

9.6   External links

- Blastocyst transfer and fertility treatment

- Risks of blastocyst transfer

- Blastocyst photos at different stages of development

- Diagram at weber.edu

- Blastocyst Differentiation Diagram

Chapter 10

Blastula

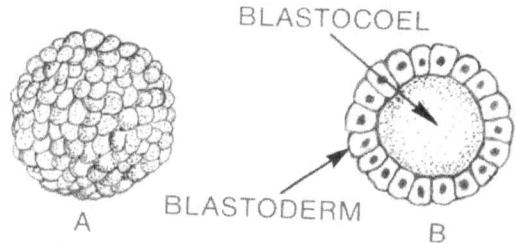

Blastocoel and blastoderm

The **blastula** (from Greek βλαστός (blastos), meaning "sprout") is a hollow sphere of cells, referred to as blastomeres, surrounding an inner fluid-filled cavity called the blastocoele formed during an early stage of embryonic development in animals.[1] Embryo development begins with a sperm fertilizing an egg to become a zygote which undergoes many cleavages to develop into a ball of cells called a morula. Only when the blastocoele is formed does the early embryo become a blastula. The blastula precedes the formation of the gastrula in which the germ layers of the embryo form.[2]

A common feature of a vertebrate blastula is that it consists of a layer of blastomeres, known as the blastoderm, which surrounds the blastocoele.[3][4] In mammals the blastula is referred to as a blastocyst. The blastocyst contains an embryoblast (or inner cell mass) that will eventually give rise to the definitive structures of the fetus, and the trophoblast, which goes on to form the extra-embryonic tissues.[2][5]

During the blastula stage of development, a significant amount of activity occurs within the early embryo to establish cell polarity, cell specification, axis formation, and regulate gene expression.[6] In many animals such as *Drosophila* and *Xenopus*, the mid blastula transition (MBT) is a crucial step in development during which the maternal mRNA is degraded and control over development is passed to the embryo.[7] Many of the interactions between blastomeres are dependent on cadherin expression, particularly E-cadherin in mammals and EP-cadherin in amphibians.[6]

The study of the blastula and of cell specification has many implications on the field of stem cell research as well as the continued improvement of fertility treatments.[5] Embryonic stem cells are a field which, though controversial, have tremendous potential for treating disease. In *Xenopus*, blastomeres behave as pluripotent stem cells which can migrate down several pathways, depending on cell signaling.[8] By manipulating the cell signals during the blastula stage of development, various tissues can be formed. This potential can be instrumental in regenerative medicine for disease and injury cases. In vitro fertilisation involves implantation of a blastula into a mother's uterus.[9] Blastula cell implantation could potentially serve to eliminate infertility.

10.1 Development

The blastula stage of early embryo development begins with the appearance of the blastocoele. The origin of the blastocoele in *Xenopus* has been shown to be from the first cleavage furrow, which is widened and sealed with tight junctions to create a cavity.[10]

In many organisms the development of the embryo up to this point and for the early part of the blastula stage is controlled by maternal mRNA, so called because it was produced in the egg prior to fertilization and is therefore exclusively from the mother.[11][12]

10.1.1 Mid-blastula transition

In many organisms including *Xenopus* and *Drosophila*, the mid-blastula transition usually occurs after a particular number of cell divisions for a given species, and is defined by the ending of the synchronous cell division cycles of the early blastula development, and the lengthening of the cell cycles by the addition of the G1 and G2 phases. Prior to this transition, cleavage occurs with only the synthesis and mitosis phases of the cell cycle.[12] The addition of the two growth phases into the cell cycle allows for the cells to increase in size, as up to this point the blastomeres undergo reductive divisions in which the overall size of the embryo does not increase, but more cells are created. This transition begins the growth in size of the organism.[2]

The mid-blastula transition is also characterised by a marked increase in transcription of new, non-maternal mRNA transcribed from the genome of the organism. Large amounts of the maternal mRNA are destroyed at this point, either by proteins such as SMAUG in *Drosophila*[13] or by microRNA.[14] These two processes shift the control of the embryo from the maternal mRNA to the nuclei.

10.2 Structure

A blastula is a sphere of cells surrounding a blastocoele. The blastocoele is a fluid filled cavity which contains amino acids, proteins, growth factors, sugars, ions and other components which are necessary for cellular differentiation. The blastocoele also allows blastomeres to move during the process of gastrulation.[15]

In *Xenopus* embryos, the blastula is composed of three different regions. The animal cap forms the roof of the blastocoele and goes on primarily to form ectodermal derivatives. The equatorial or marginal zone, which compose the walls of the blastocoel differentiate primarily into mesodermal tissue. The vegetal mass is composed of the blastocoel floor and primarily develops into endodermal tissue.[6]

In the mammalian blastocyst (term for mammalian blastula) there are three lineages that give rise to later tissue development. The epiblast gives rise to the fetus itself while the trophoblast develops into part of the placenta and the primitive endoderm becomes the yolk sac.[5]

In mouse embryo, blastocoele formation begins at the 32-cell stage. During this process, water enters the embryo, aided by an osmotic gradient which is the result of Na^+/K^+ ATPases that produce a high Na^+ gradient on the basolateral side of the trophectoderm. This movement of water is facilitated by aquaporins. A seal is created by tight junctions of the epithelial cells that line the blastocoele.[5]

10.2.1 Cellular adhesion

Tight junctions are very important in embryo development. In the blastula, these cadherin mediated cell interactions are essential to development of epithelium which are most important to paracellular transport, maintenance of cell polarity and the creation of a permeability seal to regulate blastocoel formation. These tight junctions arise after the polarity of epithelial cells is established which sets the foundation for further development and specification. Within the blastula, inner blastomeres are generally non-polar while epithelial cells demonstrate polarity.[15]

Mammalian embryos undergo compaction around the 8-cell stage where E-cadherins as well as alpha and beta catenins are expressed. This process makes a ball of embryonic cells which are capable of interacting, rather than a group of diffuse and undifferentiated cells. E-cadherin adhesion defines the apico-basal axis in the developing embryo and turns the embryo from an indistinct ball of cells to a more polarized phenotype which sets the stage for further development into a fully formed blastocyst.[15]

Xenopus membrane polarity is established with the first cell cleavage. Amphibian EP-cadherin and XB/U cadherin perform a similar role as E-cadherin in mammals establishing blastomere polarity and solidifying cell-cell interactions which are crucial for further development.[15]

10.3 Clinical implications

10.3.1 Fertilization technologies

Experiments with implantation in mice show that hormonal induction, superovulation and artificial insemination successfully produce preimplantion mice embryos. In the mice, ninety percent of the females were induced by mechanical stimulation to undergo pregnancy and implant at least one embryo.[16] These results prove to be encouraging because they provide a basis for potential implantation in other mammalian species, such as humans.

10.3.2 Stem cells

Blastula-stage cells can behave as pluripotent stem cells in many species. Pluripotent stem cells are the starting point to produce organ specific cells that can potentially aid in repair and prevention of injury and degeneration. Combining the expression of transcription factors and locational positioning of the blastula cells can lead to the development of induced functional organs and tissues. Pluripotent *Xenopus* cells, when used in an in vivo strategy, were able to form into functional retinas. By transplanting them to the eye field on the neural plate, and by inducing several misexpressions of transcription factors, the cells were committed to the retinal lineage and could guide vision based behavior in the *Xenopus*.[17]

10.4 See also

- Blastocyst

- Cellular differentiation

- Gastrulation

- Polarity in embryogenesis

10.5 Notes and references

[1] "Blastula". *Encyclopedia Britannica*. 2013.

[2] Gilbert, Scott (2010). *Developmental Biology 9th Ed + Devbio Labortatory Vade Mecum3*. Sinauer Associates Inc. pp. 243–247, 161. ISBN 978-0-87893-558-1.

[3] Lombardi, Julian (1998). "Embryogenesis". *Comparative vertebrate reproduction*. Springer. p. 226. ISBN 978-0-7923-8336-9.

[4] Forgács & Newman, 2005: p. 27

[5] Cockburn, Katie; Rossant, Janet (1 April 2010). "Making the blastocyst: lessons from the mouse". *Journal of Clinical Investigation* **120** (4): 995–1003. doi:10.1172/JCI41229.

[6] Heasman, J (November 1997). "Patterning the *Xenopus* blastula". *Development (Cambridge, England)* **124** (21): 4179–91. PMID 9334267.

[7] Tadros, Wael; Lipshitz, Howard D. (1 March 2004). "Setting the stage for development: mRNA translation and stability during oocyte maturation and egg activation in *Drosophila*". *Developmental Dynamics* **232** (3): 593–608. doi:10.1002/dvdy.20297. PMID 15704150.

[8] Gourdon, John B.; Standley, Henrietta J. (Dec 2002). "Uncommitted *Xenopus* blastula cells can be directed to uniform muscle gene expression by gradient interpretation and a community effect". *The International Journal of Developmental Biology (Cambridge, UK)* **46** (8): 993–8. PMID 12533022.

[9] Toth, Attila. "Treatment: Addressing the Causes of Infertility in Men and Women". Macleod Laboratory. Retrieved 22 March 2013.

[10] Kalt, MR (August 1971). "The relationship between cleavage and blastocoel formation in *Xenopus laevis*. I. Light microscopic observations.". *Journal of embryology and experimental morphology* **26** (1): 37–49. PMID 5565077.

[11] Tadros, W; Lipshitz, HD (March 2005). "Setting the stage for development: mRNA translation and stability during oocyte maturation and egg activation in *Drosophila*". *Developmental Dynamics* **232** (3): 593–608. doi:10.1002/dvdy.20297. PMID 15704150.

[12] Etkin, LD (1988). "Regulation of the mid-blastula transition in amphibians.". *Developmental Biology* **5**: 209–25. doi:10.1007/978-1-4615-6817-9_7. PMID 3077975.

[13] Tadros, W; Westwood, JT; Lipshitz, HD (June 2007). "The mother-to-child transition.". *Developmental Cell* **12** (6): 847–9. doi:10.1016/j.devcel.2007.05.009. PMID 17543857.

[14] Weigel, D; Izaurralde, E (24 March 2006). "A tiny helper lightens the maternal load.". *Cell* **124** (6): 1117–8. doi:10.1016/j.cell.2006.03.005. PMID 16564001.

[15] Fleming, Tom P.; Papenbrock, Tom; Fesenko, Irina; Hausen, Peter; Sheth, Bhavwanti (1 August 2000). "Assembly of tight junctions during early vertebrate development". *Seminars in Cell & Developmental Biology* **11** (4): 291–299. doi:10.1006/scdb.2000.0179. PMID 10966863.

[16] Watson, J.G. (Oct 1977). "Collection and Transfer of Preimplantation Mouse Embryos". *Biology of Reproduction* **17** (3): 453–8. doi:10.1095/biolreprod17.3.453. PMID 901897.

[17] Viczian, Andrea S.; Solessio, Eduardo C.; Lyou, Yung; Zuber, Michael E (Aug 2009). "Generation of Functional Eyes from Pluripotent Cells". *PLoS Biology* **7** (8): e1000174. doi:10.1371/journal.pbio.1000174. PMID 19688031.

10.6 Bibliography

- Forgács, G. & Newman, Stuart A. (2005). "Cleavage and blastula formation". *Biological physics of the developing embryo*. Cambridge University Press. ISBN 978-0-521-78337-8.

- Cullen, K.E. (2009). "embryology and early animal development". *Encyclopedia of life science, Volume 2*. Infobase. ISBN 978-0-8160-7008-4.

- McGeady, Thomas A., ed. (2006). "Gastrulation". *Veterinary embryology*. Wiley-Blackwell. ISBN 978-1-4051-1147-8.

Chapter 11

Inner cell mass

In early embryogenesis of most eutherian mammals, the **inner cell mass** (abbreviated **ICM** and also known as the **embryoblast** or pluriblast, the latter term being applicable to all mammals) is the mass of cells inside the primordial embryo that will eventually give rise to the definitive structures of the fetus. This structure forms in the earliest steps of development, before implantation into the endometrium of the uterus has occurred. The ICM lies within the blastocoele (more correctly termed "blastocyst cavity," as it is not strictly homologous to the blastocoele of anamniote vertebrates) and is entirely surrounded by the single layer of cells called trophoblast.

11.1 Further development

The physical and functional separation of the inner cell mass from the trophectoderm (TE) is a special feature of mammalian development and is the first cell lineage specification in these embryos. Following fertilization in the oviduct, the mammalian embryo undergoes a relatively slow round of cleavages to produce an eight cell morula. Each cell of the morula, called a blastomere, increases surface contact with its neighbors in a process called compaction. This results in a polarization of the cells within the morula, and further cleavage yields a blastocyst of roughly 32 cells.[1] In mice, about 12 internal cells comprise the new inner cell mass and 20 – 24 cells comprise the surrounding trophectoderm.[2][3] There is variation between species of mammals as to number of cells at compaction with bovine embryos showing differences related to compaction as early as 9-15 cells and in rabbits not until after 32 cells.[4] There is also interspecies variation in gene expression patterns in early embryos [5]

The ICM and the TE will generate distinctly different cell types as implantation starts and embryogenesis continues. Trophectoderm cells form extraembryonic tissues, which act in a supporting role for the embryo proper. Furthermore, these cells pump fluid into the interior of the blastocyst, causing the formation of a polarized blastocyst with the ICM attached to the trophectoderm at one end (see figure). This difference in cellular localization causes the ICM cells exposed to the fluid cavity to adopt a primitive endoderm (or hypoblast) fate, while the remaining cells adopt a primitive ectoderm (or epiblast) fate. The hypoblast contributes to extraembryonic membranes and the epiblast will give rise to the ultimate embryo proper as well as some extraembryonic tissues.[1]

11.2 Regulation of cellular specification

Since segregation of pluripotent cells of the inner cell mass from the remainder of the blastocyst is integral to mammalian development, considerable research has been performed to elucidate the corresponding cellular and molecular mechanisms of this process. There is primary interest in which transcription factors and signaling molecules direct blastomere asymmetric divisions leading to what are known as inside and outside cells and thus cell lineage specification. However, due to the variability and regulative nature of mammalian embryos, experimental evidence for establishing these early fates remains incomplete.[2]

At the transcription level, the transcription factors Oct4, Nanog, Cdx2, and Tead4 have all been implicated in establishing and reinforcing the specification of the ICM and the TE in early mouse embryos.[2]

- Oct4: *Oct4* is expressed in the ICM and participate in maintaining its pluripotency, a role that has been recapitulated in ICM derived mouse embryonic stem cells.[6] *Oct4* genetic knockout cells both in vivo and in culture display TE morphological characteristics. It has been shown that one transcriptional target of Oct4 is the *Fgf4* gene. This gene normally encodes a ligand secreted by the ICM, which induces proliferation in the adjacent polar TE.[6]

- Nanog: *Nanog* is also expressed in the ICM and par-

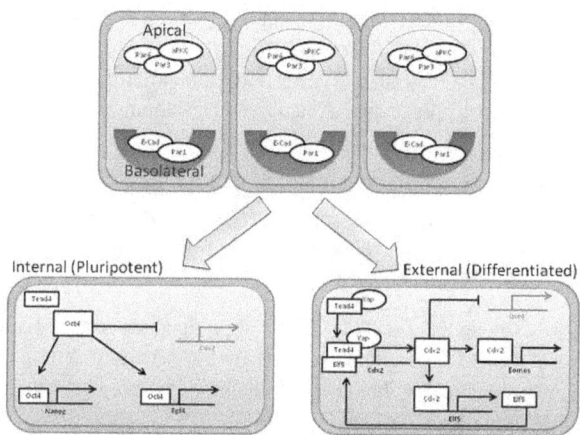

Early embryo apical and basolateral polarization is established at the 8-16 cell stage following compaction. This initial difference in environment strengthens a transcriptional feedback loop in either an internal or external direction. Inside cells express high levels of Oct4 *which maintains pluripotency and suppresses* Cdx2. *Outside cells express high levels of* Cdx2 *which causes TE differentiation and suppresses* Oct4.

ticipates in maintaining its pluripotency. In contrast with *Oct4*, studies of *Nanog*-null mice do not show the reversion of the ICM to a TE-like morphology, but demonstrate that loss of *Nanog* prevents the ICM from generating primitive endoderm.[7]

- Cdx2: *Cdx2* is strongly expressed in the TE and is required for maintaining its specification. Knockout mice for the *Cdx2* gene undergo compaction, but lose the TE epithelial integrity during the late blastocyst stage. Furthermore, *Oct4* expression is subsequently raised in these TE cells, indicating Cdx2 plays a role in suppressing *Oct4* in this cell lineage. Moreover, embryonic stem cells can be generated from *Cdx2*-null mice, demonstrating that Cdx2 is not essential for ICM specification.[8]

- Tead4: Like *Cdx2*, *Tead4* is required for TE function, although the transcription factor is expressed ubiquitously. *Tead4*-null mice similarly undergo compaction, but fail to generate the blastocoel cavity. Like *Cdx2*-null embryos, the Tead4-null embryos can yield embryonic stem cells, indicating that Tead4 is dispensable for ICM specification.[9] Recent work has shown that *Tead4* may help to upregulate Cdx2 in the TE and its transcriptional activity depends on the coactivator Yap. Yap's nuclear localization in outside cells allows it to contribute to TE specificity, whereas inside cells sequester Yap in the cytoplasm through a phosphorylation event.[10]

Together these transcription factors function in a positive feedback loop that strengthens the ICM to TE cellular allocation. Initial polarization of blastomeres occurs at the 8-16 cell stage. An apical-basolateral polarity is visible through the visualization of apical markers such as Par3, Par6, and aPKC as well as the basal marker E-Cadherin.[2] The establishment of such a polarity during compaction is thought to generate an environmental identity for inside and outside cells of the embryo. Consequently, stochastic expression of the above transcription factors is amplified into a feedback loop that specifies outside cells to a TE fate and inside cells to an ICM fate. In the model, an apical environment turns on *Cdx2*, which upregulates its own expression through a downstream transcription factor, Elf5. In concert with a third transcription factor, Eomes, these genes act to suppress pluripotency genes like *Oct4* and *Nanog* in the outside cells.[2][8] Thus, TE becomes specified and differentiates. Inside cells, however, do not turn on the *Cdx2* gene, and express high levels of *Oct4*, *Nanog*, and *Sox2*,[2][3] These genes suppress *Cdx2* and the inside cells maintain pluripotency generate the ICM and eventually the rest of the embryo proper.

Although this dichotomy of genetic interactions is clearly required to divide the blastomeres of the mouse embryo into both the ICM and TE identities, the initiation of these feedback loops remains under debate. Whether they are established stochastically or through an even earlier asymmetry is unclear, and current research seeks to identify earlier markers of asymmetry. For example, some research correlates the first two cleavages during embryogenesis with respect to the prospective animal and vegetal poles with ultimate specification. The asymmetric division of epigenetic information during these first two cleavages, and the orientation and order in which they occur, may contribute to a cell's position either inside or outside the morula,.[11][12]

11.3 Stem cells

Blastomeres isolated from the ICM of mammalian embryos and grown in culture are known as embryonic stem (ES) cells. These pluripotent cells, when grown in a carefully coordinated media, can give rise to all three germ layers (ectoderm, endoderm, and mesoderm) of the adult body.[13] For example, the transcription factor LIF4 is required for mouse ES cells to be maintained in vitro.[14] Blastomeres are dissociated from an isolated ICM in an early blastocyst, and their transcriptional code governed by *Oct4*, *Sox2*, and *Nanog* helps maintain an undifferentiated state.

One benefit to the regulative nature in which mammalian embryos develop is the manipulation of blastomeres of the ICM to generate knockout mice. In mouse, mutations in a gene of interest can be introduced retrovirally into cultured ES cells, and these can be reintroduced into the ICM of an intact embryo. The result is a chimeric mouse, which develops with a portion of its cells containing the ES cell genome. The aim of such a procedure is to incorporate the mutated gene into the germ line of the mouse such that its progeny will be missing one or both alleles of the gene of interest. Geneticists widely take advantage of this ICM manipulation technique in studying the function of genes in the mammalian system,.[1][13]

11.4 Additional images

- Blastodermic vesicle of Vespertilio murinus.

- Section through embryonic disk of Vespertilio murinus.

11.5 See also

- Homeobox genes

11.6 References

[1] Wolpert, Lewis. Principles of Development: Third Edition. 2007. Oxford University Press.

[2] Marikawa, Yusuke, et al. Establishment of Trophectoderm and Inner Cell Mass Lineages in the Mouse Embryo. Molecular Reproduction & Development 76:1019–1032 (2009)

[3] Suwinska A, Czołowska R, Ozdze_nski W, Tarkowski AK. 2008. Blastomeres of the mouse embryo lose totipotency after the fifth cleavage division: Expression of Cdx2 and Oct4 and developmental potential of inner and outer blastomeres of 16- and 32-cell embryos. Dev Biol 322:133–144.

[4] Koyama *et al* Analysis of Polarity of Bovine and Rabbit Embryos by Scanning Electron Microscopy Biol of Reproduction, 50, 163-170 1994

[5] Kuijk, *et al* Validation of reference genes for quantitative RT-PCR studies in porcine oocytes and preimplantation embryos BMC Developmental Biology 2007, 7:58 doi: 10.1186/1471-213X-7-58

[6] Nichols J, Zevnik B, Anastassiadis K, Niwa H, Klewe-Nebenius D, Chambers I, Sch€oler H, Smith A. 1998. Formation of pluripotent stem cells in the mammalian embryo depends on the POU transcription factor Oct4. Cell 95:379–391.

[7] Rodda DJ, Chew JL, Lim LH, Loh YH, Wang B, Ng HH, Robson P. 2005. Transcriptional regulation of nanog by OCT4 and SOX2. J Biol Chem 280:24731–24737.

[8] Strumpf D, Mao CA, Yamanaka Y, Ralston A, Chawengsaksophak K, Beck F, Rossant J. 2005. Cdx2 is required for correct cell fate specification and differentiation of trophectoderm in the mouse blastocyst. Development 132:2093–2102.

[9] Nishioka N, Yamamoto S, Kiyonari H, Sato H, Sawada A, Ota M, Nakao K, Sasaki H. 2008. *Tead4* is required for specification of trophectoderm in pre-implantation mouse embryos. Mech Dev 125:270–283.

[10] Nishioka N, et al. 2009. The Hippo signaling pathway components Lats and Yap pattern Tead4 activity to distinguish mouse trophectoderm from inner cell mass. Dev Cell 16: 398–410.

[11] Bischoff, Marcus, et al. Formation of the embryonic-abembryonic axis of the mouse blastocyst: relationships between orientation of early cleavage divisions and pattern of symmetric/asymmetric divisions. Development 135, 953-962 (2008)

[12] Jedrusik, Agnieszka, et al. Role of Cdx2 and cell polarity in cell allocation and specification of trophectoderm and inner cell mass in the mouse embryo. Genes Dev. 2008 22: 2692-2706

[13] Robertson, Elizabeth , et al. Germ-line transmission of genes introduced into cultured pluripotential cells by retroviral vector. Nature 323, 445 - 448 (2 October 1986)

[14] Smith AG, Heath JK, Donaldson DD, Wong GG, Moreau J, Stahl M and Rogers D (1988) Inhibition of pluripotential embryonic stem cell differentiation by purified polypeptides. Nature, 336, 688–690

11.7 External links

- Thomas A. Marino, Ph.D. - Embryology Lectures, Temple university (archive)

- Week 1: Implantation

Chapter 12

Bilaminar blastocyst

Bilaminar blastocyst or **Bilaminar disc** refers to the epiblast and the hypoblast, evolved from the embryoblast.[1][2] These two layers are sandwiched between two balloons: the primitive yolk sac and the amniotic cavity.

The inner cell mass, the embryoblast, begins to transform into two distinct epithelial layers just before implantation occurs. The epiblast is the outer layer that consists of columnar cells.The inner layer is called the hypoblast, or primitive endoderm, which is composed of cuboidal cells. As the two layers become evident, a basement membrane presents itself between the layers. The final two layers of the embryoblast are known collectively as the **bilaminar embryonic disc** as well as the **bilaminar blastocyst** or **bilaminar blastoderm**. This bilaminar blastocyst also defines the primitive dorsal ventral axis.[3] Blastocyst implantation will occur during the second week of fetal development in the endometrium of the uterus;[4] the epiblast is dorsal and the hypoblast is ventral.[3]

12.1 Formation of the blastocyst

The zygote undergoes cleavage as it journeys from the oviduct to the uterus. As it transforms from 2 to 4 to 8 to 16 cells, it becomes a Morula. During these divisions, the zygote remains the same size, only the amount of cells increases. The morula differentiates into an outer and inner group of cells: the peripheral outer cell layer, the trophoblast, and the central inner cell mass, the embryoblast. The trophoblast goes on to become the fetal portion of the placenta and related extraembryonic membranes. The epiblast and hypoblast arise from the embryoblast and later give rise to the embryo proper and its affiliated extraembryonic membranes. Once the zygote has differentiated into 30 cells, it starts to form a fluid-filled central cavity called the blastocyst cavity (blastocoele).[3] This cavity is essential because as the embryo continues to divide, the outer layer of cells grows very crowded and they have a tough time gaining adequate nutrients from surrounding

fluid. Therefore, the blastocyst cavity serves as a nutrient center and the fluid is able to reach and feed cells so that they can continue growing and dividing.[4] The embryo is called a blastocyst at about the 6th day of development once it has reached nearly 100 cells. Once this has happened, the embryo begins its journey through the uterus to start implanting in the endometrium.[3]

12.2 Becoming bilaminar

The zygote first transformed into a morula through cleavage and then more divisions lead to a blastocyst that consisted of just a trophoblast, and an embryoblast. By the end of the first week, the embryoblast has begun separating into two layers: the epiblast and hypoblast also called the primitive endoderm. At the embryonic pole of the blastocyst, the amniotic cavity finds a home between the epiblast and the trophoblast. The epiblast stretches to surround the cavity very quickly and this layer of the epiblast becomes known as the amnion, which is one of the four extraembryonic membranes. The rest of the hypoblast and epiblast, not including the amnion, is what contributes to the **bilaminar embryonic disc** (**bilaminar blastoderm/blastocyst**), which sits between the amniotic cavity and the blastocyst cavity. The embryo proper and extramembryonic membranes are later derived from the embryonic disc.[3]

12.3 Establishment of the amniotic cavity

Beginning on day 8, the amniotic cavity is the first new cavity to form during the second week of development.[3] Fluid collects between the epiblast and the hypoblast, which splits the epiblast into two portions. The layer at the embryonic pole grows around the amniotic cavity, creating a barrier from the cytotrophoblast. This becomes known as the amnion, which is one of the four extraembyonic membranes and the cells it comprises are referred to as amnioblasts.[5] Although, the amniotic cavity starts off small it eventually grows to be larger than the blastocyst and by week 8, the whole embryo is encompassed by the amnion.[3]

12.4 Formation of the yolk sac and chorionic cavity

The formation of the chorionic cavity (Extra-embryonic coelom) and the yolk sac (umbilical vesicle) is still up for debate. The thought of how the yolk sac membranes are formed begins with an increase in production of hypoblast cells, succeeded by different patterns of migration. On day 8, the first portion of hypoblast cells begin their migration and make what is known as the primary yolk sac, or Heuser's membrane (exocoelomic membrane). By day 12, the primary yolk sac has been disestablished by a new batch of migrating hypoblast cells that now contribute to the definitive yolk sac.[3]

While the primary yolk sac is forming, extraembryonic mesoderm makes its way into the blastocyst cavity to fill it with loosely packed cells. When the extraembryonic mesoderm is separated into two portions, a new gap arises called the chorionic cavity, or the extra-embryonic coelom. This new cavity is responsible for detaching the embryo and its amnion and yolk sac from the far wall of the blastocyst, which is now named the chorion. When the extraembryonic mesoderm splits into two layers, the amnion, yolk sac, and chorion follow its lead and also become double layered. The chorion and amnion are composed of extraembryonic ectoderm and mesoderm, where as the yolk sac is made of extraembryonic endoderm and mesoderm. When day 13 rolls around, the connecting stalk, a dense portion of extraembryonic mesoderm, restrains the embryonic disc in the chorionic cavity.[3]

12.4.1 Yolk sac during development

Like the amnion, the yolk sac is simply an extraembryonic membrane that surrounds a cavity. Formation of the definitive yolk sac happens after the extraembryonic mesoderm splits, and it becomes a double layered structure with hypoblast-derived endoderm on the inside and mesoderm surrounding the outside. The definitive yolk sac contributes greatly to the embryo during the 4th week of development and it executes critical functions for the embryo. One of which being the formation of blood, or hematopoiesis. Also, Primordial germ cells are first found in the wall of the yolk sac. After the 4th week of development, the growing embryonic disc becomes a great deal larger than the yolk sac and its presence usually dies out before birth. However,

seldom will the yolk sac remain as deviation of the digestive tract named Meckel's diverticulum.[3]

12.5 Epiblast cells during gastrulation

The third week of development and the formation of the primitive streak sparks the beginning of gastrulation.[3] Gastrulation is when the three germ cell layers develop as well as an organism's body plan.[6] During gastrulation, cells of the epiblast, a layer of the bilaminar blastocyst, migrate towards the primitive streak, enter it, and then move apart from it through a process called ingression.[3]

12.5.1 Definitive endoderm development

On day 16, epiblast cells that are next to the primitive streak experience epithelial-to-mesenchymal transformation as they ingress through the primitive streak. The first wave of epiblast cells takes over the hypoblast, which slowly becomes replaced by new cells that eventually constitute the definitive endoderm. The definitive endoderm is what makes the lining of the gut and other associated gut structures.[3]

12.5.2 Intraembryonic mesoderm development

Also beginning on day 16, some of the ingressing epiblast cells make their way into the area between the epiblast and the newly forming definitive endoderm. This layer of cells becomes known as intraembryonic mesoderm. After the cells have moved bilaterally from the primitive streak and matured, four divisions of intraembryonic mesoderm are made; cardiogenic mesoderm, paraxial mesoderm, intermediate mesoderm and lateral plate mesoderm.[3]

12.5.3 Ectoderm development

After the definitive endoderm and intraembryonic mesoderm formations are complete, the remaining epiblast cells do not ingress through the primitive streak; rather they remain on the outside and form the ectoderm. It is not long until the ectoderm becomes the neural plate and surface ectoderm. Due to the fact that an embryo develops cranial to caudal, the formation of ectoderm does not happen at the same rate during development. The more inferior portion of the primitive streak will still have epiblast cells ingressing to make intraembryonic mesoderm, while the more superior portion has stop ingressing. However, eventually gastrulation finishes and the three germ layers are complete.[3]

12.6 References

[1] The Third Week Of Life:

[2] Bilaminar Disc

[3] Schoenwolf, Gary C., and William J. Larsen. Larsen's Human Embryology. 4th ed. Philadelphia: Churchill Livingstone/Elsevier, 2009. Print.

[4] "Bilaminar Embryonic Disc." Atlas of Human Embryology. Chronolab A.G. Switzerland, n.d. Web. 27 Nov. 2012. <http://www.embryo.chronolab.com/formation.htm>.

[5] "10.1 Early Development and Implantation." The Embryoblast. N.p., n.d. Web. 29 Nov. 2012. <http://www.embryology.ch/anglais/fplacenta/fecond04.html>

[6] http://www.gastrulation.org/

Chapter 13

Hypoblast

The **hypoblast** is a tissue type that forms from the inner cell mass.[1] It lies beneath the epiblast and consists of small cuboidal cells.[2]

Extraembryonic endoderm (including Yolk sac) is derived from hypoblast cells. The absence of hypoblast results in multiple primitive streaks in chicken embryos.[3] The formation of the primitive streak, through which gastrulation occurs, is induced by Koller's sickle.[4]

13.1 Structure

13.1.1 In mice

In mouse embryo, the visceral endoderm develop from the primitive endoderm of the blastocyst during the implantation stage covering the epiblast cells and elongates to become an egg cylinder. A distinct morphological domain has been identified by Martin and colleagues, at the distal tip of the mouse egg cylinder, thus this domain was called distal visceral domain (DVE).[5] The DVE cells will move unilaterally to the future anterior until reaching the embryonic/ extra embryonic boundary and at this point, the DVE cells are also named as anterior visceral endoderm (AVE).[6] This migration has been proved to be essential for establishing anteroposterior axis. Besides the AVE, another cell population appears to be separated at the posterior edge of the embryonic egg cylinder, referred to as posterior visceral endoderm (PVE). However, the function of this cell population was not as well studied as AVE.

13.2 Function

Although the hypoblast does not contribute to the embryo, it has great influences on the orientation of the embryonic axis. For example, the AVE in hypoblast plays important role in positioning the primitive streak at the midland of the amniote embryos. In chick, people had observed that removal of the hypoblast caused multiple, ectopic primitive streaks formation.[7] Similarly, in mice embryo, the AVE expresses secreted molecules, including two antagonists of Nodal signaling, Cerberus-like (Cerl) and a TGFβ superfamily molecule, Lefty1. It was shown that Cerberus−/−;Lefty1−/− compound mutants mice developed a primitive streak ectopically in the embryo.[8] In addition, there is also finding suggested that the hypoblast also inhibit primitive streak formation by depositing extracellular matrix components to inhibit epithelial-mesenchymal transition (EMT).[9] Besides the role of positioning the site of gastulation, AVE also showed other function including continued protection against caudalization of the early nervous system.[10] Also primitive endoderm derived yolk sac has major function in guaranteeing the proper organogenesis of the fetus and efficient exchange of nutrients, gases and wastes.

13.3 History

In mammals, the existence of primitive endoderm had been observed as early as the end of the 19th century as first recognized by Duval and Sobotta.[11][12] However, it took long time before people realized that the primitive endoderm will be replaced by definitive endoderm which will further develop into the gut tube. The first convincing experiment was conducted by Bellairs in chick embryo with the careful observation under electron and light microscopy. In his experiment, Bellairs demonstrated that there is a transitory endoderm cell layer in the chick embryo at its ventral surface before the formation of primitive streak. This layer of cell was replaced by definitive endoderm migration from the primitive streak through ingression and de-epithelialization.[13][14][15][16] Later on, more insights on primitive endoderm and definitive endoderm origin and formation have been provided in different species including rat and mouse, rhesus monkey, baboon et.al.[17][18][19][20][21]

13.4 References

[1] UNSW Embryology- Glossary H

[2] Moore, K. L., and Persaud, T. V. N. (2003). *The Developing Human: Clinically Oriented Embryology*. 7th Ed. Philadelphia: Elsevier. ISBN 0-7216-9412-8.

[3] Perea-Gomez A, Vella FD, Shawlot W, Oulad-Abdelghani M, Chazaud C, Meno C, Pfister V, Chen L, Robertson E, Hamada H, Behringer RR, Ang SL. (2002). "Nodal antagonists in the anterior visceral endoderm prevent the formation of multiple primitive streaks". *Dev Cell.* **3** (5): 745–56. doi:10.1016/S1534-5807(02)00321-0. PMID 12431380.

[4] Gilbert SF. Developmental Biology. 10th edition. Sunderland (MA): Sinauer Associates; 2014. Early Development in Birds. Print

[5] Rosenquist T. A., Martin G. R. (1995). Visceral endoderm-1 (VE-1): an antigen marker that distinguishes anterior from posterior embryonic visceral endoderm in the early post-implantation mouse embryo. Mech. Dev. 49, 117–121

[6] Thomas P., Beddington R. (1996). Anterior primitive endoderm may be responsible for patterning the anterior neural plate in the mouse embryo. Curr. Biol. 6, 1487–1496.

[7] Bertocchini F., Stern C. D. (2002). The hypoblast of the chick embryo positions the primitive streak by antagonizing nodal signaling. Dev. Cell 3, 735–744.

[8] Perea-Gomez A., Vella F. D., Shawlot W., Oulad-Abdelghani M., Chazaud C., Meno C., Pfister V., Chen L., Robertson E., Hamada H., et al. (2002).Nodal antagonists in the anterior visceral endoderm prevent the formation of multiple primitive streaks.

[9] Egea J., Erlacher C., Montanez E., Burtscher I., Yamagishi S., Hess M., Hampel F., Sanchez R., Rodriguez-Manzaneque M. T., Bosl M. R., et al. (2008). Genetic ablation of FLRT3 reveals a novel morphogenetic function for the anterior visceral endoderm in suppressing mesoderm differentiation.Genes Dev. 22, 3349–3362.

[10] Wilson S. W., Houart C. (2004). Early steps in the development of the forebrain. Dev. Cell 6, 167–181.

[11] Duval M. (1891). The rodent placenta. Third part. The placenta of the mouse and of the rat. J. Anat. Physiol. Normales et Pathol. de l'Homme et des Animaux 27, 24-73; 344-395; 515-612.

[12] Sobotta J. (1911). Die Entwicklung des Eies der Maus vom ersten Auftreten des Mesoderms an bis zur Ausbildung der Embryonalanlage und dem Auftreten der Allantois. I. Teil: Die Keimblase. Archiv. fur mikroskopische Anatomie 78, 271–352.

[13] Bellairs R. (1953a). Studies on the development of the foregut in the chick blastoderm. 1. The presumptive foregut area. J. Embryol. Exp. Morph. 1, 115–124.

[14] Bellairs R. (1953b). Studies on the development of the foregut in the chick blastoderm. 2. The morphogenetic movements. J. Embryol. Exp. Morph. 1, 369–385.

[15] Bellairs R. (1964). Biological aspects of the yolk of the hen's egg. Adv. Morphog. 4, 217–272.

[16] Bellairs R. (1986). The primitive streak. Anat. Embryol.

[17] Enders A. C., Given R. L., Schlafke S. (1978). Differentiation and migration of endoderm in the rat and mouse at implantation. Anat. Rec. 190, 65–77.

[18] Enders A. C., Schlafke S., Hendrickx A. G. (1986). Differentiation of the embryonic disc, amnion, and yolk sac in the rhesus monkey. Am. J. Anat. 177, 161–185.

[19] Enders A. C., Lantz K. C., Schlafke S. (1990). Differentiation of the inner cell mass of the baboon blastocyst. Anat. Rec. 226, 237–248.

[20] Gardner R. L. (1982). Investigation of cell lineage and differentiation in the extraembryonic endoderm of the mouse embryo. J. Embryol. Exp. Morphol. 68, 175–198.

[21] Gardner R. L. (1984). An in situ cell marker for clonal analysis of development of the extraembryonic endoderm in the mouse. J. Embryol. Exp. Morphol. 80, 251–288.

13.5 External links

- http://www.embryology.ch/allemand/iperiodembry/carnegie02.html

- http://www.med.umich.edu/lrc/coursepages/M1/embryology/embryo/04secondweek.htm

- http://isc.temple.edu/marino/embryology/EMBII97/sld005.htm

Chapter 14

Epiblast

In amniote animal embryology, the **epiblast** is one of two distinct layers arising from the inner cell mass in the mammalian blastocyst or from the blastodisc in reptiles and birds. It derives the embryo proper through its differentiation into the three primary germ layers, ectoderm, mesoderm and endoderm, during gastrulation. The amnionic ectoderm and extraembryonic mesoderm also originate from the epiblast.

14.1 Mammals

In mammalian embryogenesis, differentiation and segregation of cells composing the inner cell mass of the blastocyst yields two distinct layers—the epiblast and the hypoblast. While the cuboidal hypoblast cells delaminate ventrally, away from the embryonic pole, to line the blastocoele, the remaining cells of the inner cell mass, situated between the hypoblast and the polar trophoblast, become the epiblast and comprise columnar cells.

Upon commencement of gastrulation, the primitive streak, a visible, morphological groove, appears on the posterior epiblast and orients along the anterior-posterior embryo axis. Initiated by signals from the underlying hypoblast, formation of the primitive streak is predicated on epiblast cell migration, mediated by Nodal, from the lateral-posterior regions of the epiblast to the center midline.[1] The primitive knot is situated at the anterior end of the primitive streak and serves as the organizer for gastrulation, determining epiblast cell fate by inducing the differentiation of migrating epiblast cells during gastrulation.

During gastrulation, migrating epiblast cells undergo epithelial-mesenchymal transition in order to lose cell-cell adhesion (E-cadherin), delaminate from the epiblast layer and migrate over the dorsal surface of the epiblast then down through the primitive streak. The first wave of epiblast cells to invaginate through the primitive streak invades and displaces the hypoblast to become the embryonic endoderm. The mesoderm layer is established next as migrating epiblast cells move through the primitive streak then spread out within the space between the endoderm and remaining epiblast, which once the mesoderm layer has formed ultimately becomes the definitive ectoderm. The process of gastrulation results in a trilaminar germ disc, consisting of the ectoderm, mesoderm and endoderm layers.

14.1.1 Epiblast diversity

Epiblasts exhibit diverse structure across species as a result of early embryo morphogenesis. The human epiblast assumes a disc shape, conforming to the embryonic disc morphology; whereas, the mouse epiblast develops in a cup shape within the cylindrical embryo.

14.2 Birds

Gastrulation occurs in the epiblast of avian embryos. A local thickening of the epiblast, known as Koller's sickle, is key in inducing the primitive streak, the structure through which gastrulation occurs.[2]

14.3 See also

- Embryogenesis
- Human embryogenesis
- Hypoblast

14.4 References

[1] Shen MM. Nodal signaling: developmental roles and regulation. Development 2007; 134(6): 1023-1034.

[2] Gilbert SF. Developmental Biology. 10th edition. Sunderland (MA): Sinauer Associates; 2014. Early Development in Birds. Print

14.5   External links

- http://www.embryology.ch/allemand/iperiodembry/carnegie02.html

- http://www.med.umich.edu/lrc/coursepages/M1/embryology/embryo/04secondweek.htm

- http://isc.temple.edu/marino/embryology/EMBII97/sld005.htm

Chapter 15

Trilaminar blastocyst

A **trilaminar embryo** (or **trilaminary blastoderm**, or **trilaminar germ disk**) is an early stage in the development of triploblastic organisms, which include humans and many other animals.

It is an embryo which exists as three different germ layers - the ectoderm, the mesoderm and the endoderm. These layers are arranged on top of each other like a stack of paper, giving rise to the name *trilaminar*, or "three-layered".

These three layers arise early in the third week (after gastrulation) from the epiblast (a portion of the mammalian inner cell mass).

15.1 External links

- Swiss embryology (from UL, UB, and UF) *hdisqueembry/triderm01*

- Embryology at UNSW *Notes/week3_4*

- Overview at edu.mt

Chapter 16

Germ layer

See also: Germ cell

A **germ layer** is a primary layer of cells that form during embryogenesis.[1] The three germ layers in vertebrates are particularly pronounced; however, all eumetazoans, (animals more complex than the sponge) produce two or three primary germ layers. Animals with radial symmetry, like cnidarians, produce two germ layers (the ectoderm and endoderm) making them diploblastic. Animals with bilateral symmetry produce a third layer (the mesoderm), between these two layers. making them triploblastic. Germ layers eventually give rise to all of an animal's tissues and organs through the process of organogenesis.

16.1 Germ layers

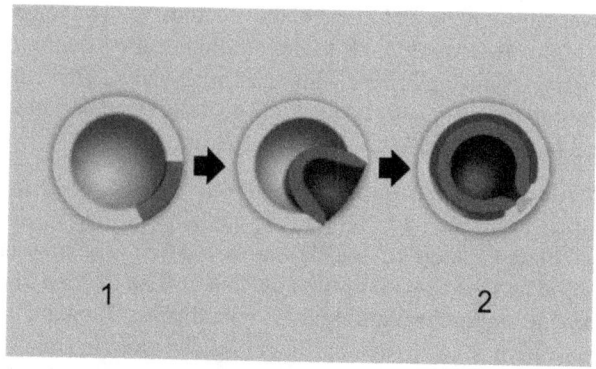

Gastrulation of a diploblast: *The formation of germ layers from a (1) blastula to a (2) gastrula. Some of the ectoderm cells (orange) move inward forming the endoderm (red).*

Caspar Friedrich Wolff observed organization of the early embryo in leaf-like layers. In 1817, Heinz Christian Pander discovered three primordial germ layers while studying chick embryos. Between 1850 and 1855, Robert Remak had further refined the germ cell layer concept, and introduced into English were the terms "mesoderm" by Huxley

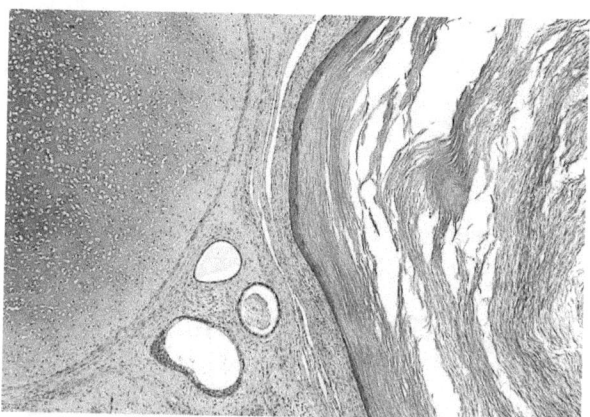

*Micrograph of a teratoma, a tumour that characteristically has tissue from all three **germ layers**. The image shows tissue derived from the mesoderm (immature cartilage - left-upper corner of image), endoderm (gastrointestinal glands - center-bottom of image) and ectoderm (epidermis - right of image). H&E stain.*

in 1871 and "ectoderm" and "endoderm" by Lankester in 1873.

Among animals, sponges show the simplest organization, having a single germ layer. Although they have differentiated cells (e.g. collar cells), they lack true tissue coordination. Diploblastic animals, Cnidaria and Ctenophora, show an increase in complexity, having two germ layers, the endoderm and ectoderm. Diploblastic animals are organized into recognisable tissues. All higher animals (from flatworms to humans) are triploblastic, possessing a mesoderm in addition to the germ layers found in Diploblasts. Triploblastic animals develop recognisable organs.

16.1.1 Development

Fertilization leads to the formation of a zygote. During the next stage, cleavage, mitotic cell divisions transform the zygote into a hollow ball of cells, a blastula. This early embryonic form undergoes gastrulation, forming a gastrula with either two or three layers (the germ layers). In all vertebrates, these progenitor cells differentiate into all adult tissues and organs.[2]

In humans, after about three days, the zygote forms a solid mass of cells by mitotic division, called a morula. This then changes to a blastocyst, consisting of an outer layer called a trophoblast, and an inner cell mass called the embryoblast. Filled with uterine fluid, the blastocyst breaks out of the zona pellucida and undergoes implantation. The inner cell mass initially has two layers: the hypoblast and epiblast. At the end of the second week, a primitive streak appears. The epiblast in this region moves towards the primitive streak, dives down into it, and forms a new layer, called the endoderm, pushing the hypoblast out of the way (this goes on to form the amnion.) The epiblast keeps moving and forms a second layer, the mesoderm. The top layer is now called the ectoderm.[3]

16.1.2   Endoderm

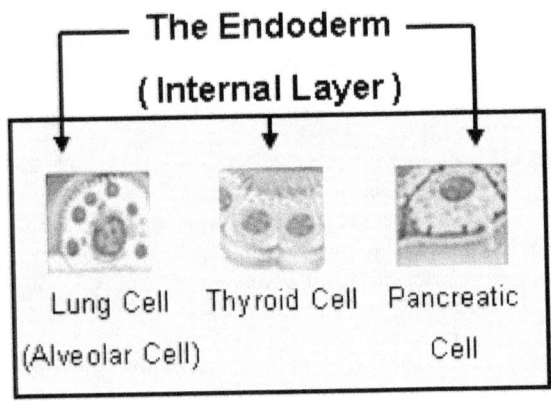

The endoderm produces tissue within the lungs, thyroid, and pancreas.

Main article: Endoderm

The **endoderm** is one of the germ layers formed during animal embryogenesis. Cells migrating inward along the archenteron form the inner layer of the gastrula, which develops into the endoderm.

The endoderm consists at first of flattened cells, which subsequently become columnar. It forms the epithelial lining of the whole of the digestive tube except part of the mouth and pharynx and the terminal part of the rectum (which are lined by involutions of the ectoderm). It also forms the lining cells of all the glands which open into the digestive tube, including those of the liver and pancreas; the epithelium of the auditory tube and tympanic cavity; the trachea, bronchi,

and air cells of the lungs; the urinary bladder and part of the urethra; and the follicle lining of the thyroid gland and thymus.

The endoderm forms: the stomach, the colon, the liver, the pancreas, the urinary bladder, the epithelial parts of trachea, the lungs, the pharynx, the thyroid, the parathyroid, and the intestines.

16.1.3   Mesoderm

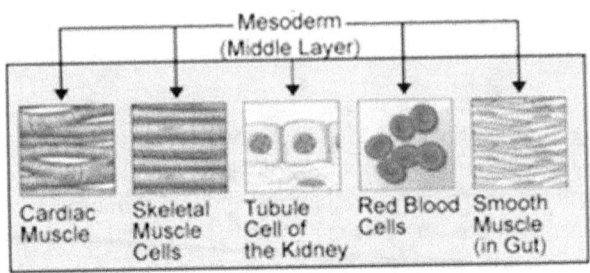

The mesoderm aids in the production of cardiac muscle, skeletal muscle, smooth muscle, tissues within the kidneys, and red blood cells.

Main article: Mesoderm

The **mesoderm** germ layer forms in the embryos of triploblastic animals. During gastrulation, some of the cells migrating inward contribute to the mesoderm, an additional layer between the endoderm and the ectoderm. The formation of a mesoderm leads to the development of a coelom. Organs formed inside a coelom can freely move, grow, and develop independently of the body wall while fluid cushions and protects them from shocks.

The mesoderm has several components which develop into tissues: intermediate mesoderm, paraxial mesoderm, lateral plate mesoderm, and chorda-mesoderm. The chorda-mesoderm develops into the notochord. The intermediate mesoderm develops into kidneys and gonads. The paraxial mesoderm develops into cartilage, skeletal muscle, and dermis. The lateral plate mesoderm develops into the circulatory system (including the heart and spleen), the wall of the gut, and wall of the human body.[4]

Through cell signaling cascades and interactions with the ectodermal and endodermal cells, the mesodermal cells begin the process of differentiation.[5]

The mesoderm forms: muscle (smooth and striated), bone, cartilage, connective tissue, adipose tissue, circulatory system, lymphatic system, dermis, genitourinary system, serous membranes, and notochord.

16.1.4 Ectoderm

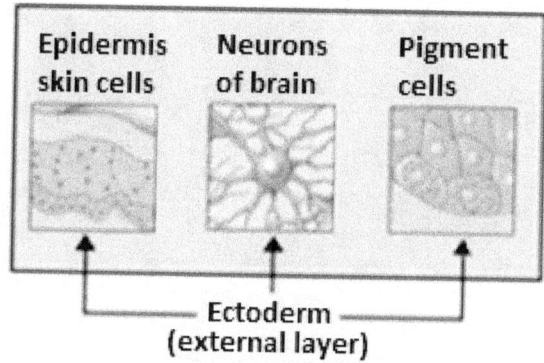

The ectoderm produces tissues within the epidermis, aids in the formation of neurons within the brain, and constructs melanocytes.

Main article: Ectoderm

The **ectoderm** generates the outer layer of the embryo, and it forms from the embryo's epiblast.[6] The ectoderm develops into the surface ectoderm, neural crest, and the neural tube.[7]

The surface ectoderm develops into: epidermis, hair, nails, lens of the eye, sebaceous glands, cornea, tooth enamel, the epithelium of the mouth and nose.

The neural crest of the ectoderm develops into: peripheral nervous system, adrenal medulla, melanocytes, facial cartilage, dentin of teeth.

The neural tube of the ectoderm develops into: brain, spinal cord, posterior pituitary, motor neurons, retina.

Note: The anterior pituitary develops from the ectodermal tissue of Rathke's pouch.

16.1.5 Neural crest

Because of its great importance, the neural crest is sometimes considered a fourth germ layer.[8] It is, however, derived from the ectoderm.

16.2 See also

- Histogenesis

- Neurulation

16.3 References

[1] Gilbert, Scott F (2003). "The Epidermis and the Origin of Cutaneous Structures". *Developmental Biology*. Sinauer Associates.

[2] Gilbert, Scott F (2000). "Comparative Embryology". *Developmental Biology*. Sinauer Associates.

[3] Gilbert, Scott F (2000). "Early Mammalian Development". *Developmental Biology*. Sinauer Associates.

[4] Gilbert, Scott F (2003). "Paraxial and Intermediate Mesoderm". *Developmental Biology*. Sinauer Associates.

[5] Brand, Thomas (1 June 2003). "Heart development: molecular insights into cardiac specification and early morphogenesis". *Developmental Biology* **258** (1): 1–19. doi:10.1016/S0012-1606(03)00112-X.

[6] Gilbert, Scott F (2003). "Early Mammalian Development". *Developmental Biology*. Sinauer Associates.

[7] Gilbert, Scott F (2003). "The Central Nervous System and The Epidermis". *Developmental Biology*. Sinauer Associates.

[8] Hall BK (2000). "The neural crest as a fourth germ layer and vertebrates as quadroblastic not triploblastic". *Evolution & Development 2, 3-5* **2**: 3–5. doi:10.1046/j.1525-142x.2000.00032.x. PMID 11256415.

Chapter 17

Archenteron

The primary gut that forms during gastrulation in the developing zygote is known as the **archenteron** or the **digestive tube**. It develops into the endoderm and mesoderm of an animal.

17.1 Formation of the Archenteron in Sea Urchins

See Gastrulation.

As primary mesenchyme cells detach from the vegetal pole in the gastrula and enter the fluid filled cavity in the center (the blastocoel), the remaining cells at the vegetal pole flatten to form a vegetal plate. This buckles inwards towards the blastocoel in a process called invagination. The cells continue to be rearranged until the shallow dip formed by invagination transforms into a deeper, narrower pouch formed by the gastrula's endoderm. This narrowing and lengthening of the archenteron is driven by convergent extension. The open end of the archenteron is called the blastopore.

The filopodia--thin fibers formed by the mesenchyme cells--found in a late gastrula contract to drag the tip of the archenteron across the blastocoel. The endoderm of the archenteron will fuse with the ectoderm of the blastocoel wall. At this point gastrulation is complete, and the gastrula has a functional digestive tube.

17.2 Similar Formation of the Archenteron in Other Animals

The indentation that is actually formed is called the lip of the blastopore or the dorsal lip in amphibians and fish, and the primitive streak in birds and mammals. Each is controlled by the dorsal blastopore, and primitive node (also known as Hensen's node), respectively.

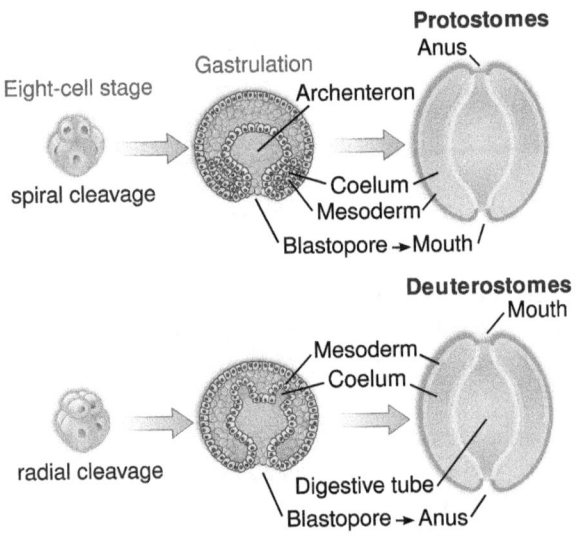

The Archenteron is labeled as the digestive tube

During Gastrulation, the Archenteron develops into the digestive tube, with the blastopore developing into either the mouth (Protostome) or the anus (Deuterostome)

17.3 External links

- Diagram

Chapter 18

Primitive streak

The **primitive streak** is a structure that forms in the blastula during the early stages of avian, reptilian and mammalian embryonic development. It forms on the dorsal (back) face of the developing embryo, toward the caudal or posterior end.

The presence of the primitive streak will establish bilateral symmetry, determine the site of gastrulation and initiate germ layer formation. To form the streak, reptiles, birds and mammals arrange mesenchymal cells along the prospective midline, establishing the second embryonic axis, as well as the place where cells will ingress and migrate during the process of gastrulation and germ layer formation.[1] The primitive streak extends through this midline and creates the left–right and cranial–caudal body axes,[2] and marks the beginning of gastrulation.[3] This process involves the ingression of mesoderm progenitors and their migration to their ultimate position,[2][4] where they will differentiate into the mesoderm germ layer[1] that, together with endoderm and ectoderm germ layers, will give rise to all the tissues of the adult organism.

18.1   Components

Given that the chicken embryo can be easily manipulated, most of our knowledge about the primitive streak comes from avian studies. The marginal zone of a chick embryo contains cells that will contribute to the streak.[4] This region has a defined anterior-to-posterior gradient in its ability to induce the primitive streak, with the posterior end having the highest potential.[5]

The epiblast, a single epithelial layer blastodisc, is the source of all embryonic material in amniotes[1] and some of its cells will give rise to the primitive streak.[4] All cells in the epiblast can respond to signals from the marginal zone,[1] but once a given region is induced by these signals and undergoes streak formation, the remaining cells in the epiblast are no longer responsive to these inductive signals and prevent the formation of another streak.[5]

Underlying the epiblast is the hypoblast, where the extra-embryonic tissue originates.[4] In the chick, the absence of the hypoblast results in multiple streaks,[6] suggesting that its presence is important for regulating the formation of a single primitive streak. In mice, this structure is known as the Anterior Visceral Endoderm (AVE).[6]

18.2   Cellular movements

The formation of the primitive streak in the blastocyst involves the coordinated movement and re-arrangement of cells in the epiblast. Even before the streak is visible, epiblast cells have started to move.[7] Two counter-rotating flows of cells meet at the posterior end, where the streak forms.[7] There is little movement in the center of these flows, while the greatest movement is observed at the periphery of the vortices.[3] The Polonaise Movement is key for the formation of the primitive streak. Cells overlaying Koller's Sickle in the posterior end of the embryo move towards the midline, meet and change direction towards the center of the epiblast. Cells from the lateral posterior marginal zone replace those cells that left Koller's Sickle by meeting at the center of this region, changing direction and extending anteriorly.[4][8] As these cells move and concentrate at the posterior end of the embryo, the streak undergoes a single- to multi-layered epithelial sheet transition that makes it a macroscopically visible structure.[4] Several mechanisms, including oriented cell division, cell-cell intercalation and chemotactic cell movement,[4] have been proposed to explain the nature of the cellular movements required to form the primitive streak.

18.3   Formation

The formation of the primitive streak relies on a complex network of signaling pathways that work together to ensure that this process is highly regulated. Activation of various secreted factors (Vg1, Nodal, Wnt8C, FGF8 and Chordin) and transcription factors (Brachyury and Goosecoid) adjacent to the site of streak formation is required for this process.[9][10][11][12][13] In addition, structures such as the hypoblast also play an important role in the regulation of streak formation. Removal of the hypoblast in the chick results in correctly patterned ectopic streaks, suggesting that the hypoblast serves to inhibit formation of the primitive streak.[13]

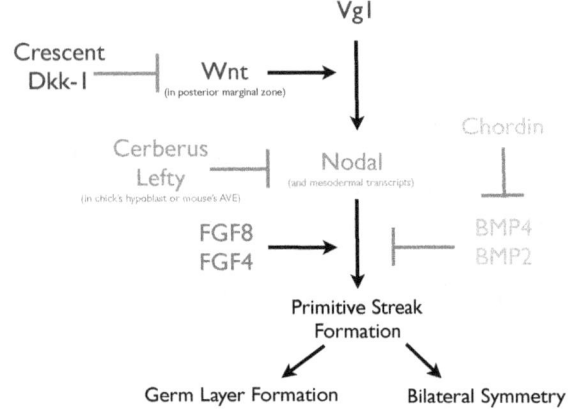

An intricate network of signaling pathways regulate the formation of the primitive streak.

18.3.1   Vg1 and Wnt signaling

Similarly, Vg1 (a TGFB family member) misexpression and grafts of the posterior marginal zone[5] in chicks can also induce ectopic streaks, but only within the marginal zone of the embryo,[11] indicating a specific characteristic of this region in its ability to induce streak formation. Several lines of evidence point to Wnt expression as the determinant of this ability. Deletion of Wnt3 in mouse embryos results in the absence of a streak formation, similarly to the phenotype of B-catenin mutant embryos.[14] In addition, mutating the intracellular negative regulator of Wnt signaling, Axin,[15] and misexpression of the chick cWnt8C[16] produces multiple streaks in mouse embryos. Localization of Wnt and components of its pathway, Lef1 and B-catenin, further supports streak-inducing role in the marginal zone.[11] Furthermore, it is expressed as a gradient decreasing from posterior to anterior,[11][12] corresponding to the streak-inducing ability of the marginal zone. Misexpression of Vg1 or Wnt1 alone failed to induce an ectopic streak in the chick, but together their misexpression resulted in ectopic streak formation, confirming that the streak-inducing ability of the posterior marginal zone could be attributed to Wnt signaling[11] and that Vg1 and Wnt must cooperate to induce this process. Misexpression of Vg1 along with Wnt antagonists, Crescent or Dkk-1, pre-

vents the formation of ectopic streaks,[11] demonstrating the importance of Wnt activity in the formation of Vg1-induced ectopic streaks and hence its implication in normal primitive streak formation.

18.3.2   Hypoblast

Any given slice from the blastoderm is able to generate a complete axis[17] until the time of gastrulation and primitive streak formation.[13] This ability to generate a streak from the pre-streak stage chick embryo[18] indicates that there must be a mechanism to ensure that only a single streak forms. The hypoblast secretes an antagonist of Nodal that prevents ectopic streak formation in the chick.[13]

18.3.3   Nodal signaling

Nodal, a known mesodermal inducer of the TGFB superfamily,[18] has been implicated in streak formation. Mouse embryos mutant for Nodal fail to gastrulate and lack most mesoderm,[19] but more than playing a role in mesoderm induction, Nodal regulates the induction and/or maintenance of the primitive streak.[19] In the presence of hypoblast, Nodal is unable to induce ectopic streaks in the chick embryo, while its removal, induces expression of Nodal, Chordin and Brachyury,[13] suggesting that the hypoblast must have a certain inhibitory effect on Nodal signaling. Indeed, the multifunctional antagonist of Nodal, Wnt and BMP signaling, Cerberus (produced in the hypoblast) and Cerberus-Short (which inhibits only Nodal), through its effect on Nodal signaling, inhibits streak formation.[13] Eventually, the hypoblast gets displaced anteriorly by the moving endoblast, allowing streak formation at the posterior end. At the anterior end, the presence of the hypoblast and the antagonists it secretes, such as Cerberus, inhibit the expression of Nodal and hence restrict streak formation to the posterior end only.[13] Similarly to the hypoblast in chick, the AVE in the mouse secretes two antagonists of Nodal signaling, Cerberus-like, Cerl, and Lefty1.[13][20] In mouse, Cer-/-; Lefty1-/- double mutants develop multiple streaks[6] as indicated by ectopic expression of Brachyury and can be partially rescued by the removal of one copy of the Nodal gene.[6] In the mouse, the AVE restricts streak formation through the redundant functions of Cer1 and Lefty1, which negatively regulate Nodal signaling.[6] The role of the mouse's AVE in ensuring the formation of a single primitive streak is evolutionarily conserved in the hypoblast of the chick.[6][13]

18.3.4 FGF signaling

Another important pathway in modulating formation of the primitive streak is FGF, which is thought to work together with Nodal to regulate this process.[18] Inhibition of FGF signaling through expression of a dominant negative receptor, using a FGF receptor inhibitor (SU5402) or depletion of FGF ligands, inhibit mesoderm formation[3] and this in turn, inhibits streak formation.[4] Furthermore, ectopic streak formation induced by Vg1 required FGF signaling.[18]

18.3.5 BMP signaling

Finally, BMP signaling is also important for regulating the process of streak formation in the chick embryo. The site of streak formation is characterized by low BMP signals, while the rest of the epiblast displays high levels of BMP activation.[21] In addition, misexpression of either BMP4 or BMP7 prevents streak formation, while the BMP inhibitor Chordin induces ectopic streak formation in the chick,[22] suggesting that streak formation is likely to require BMP inhibition.

18.4 Ethical implications

The primitive streak is an important concept in bioethics, where some experts have argued that experimentation with human embryos is permissible, but only before the primitive streak develops, generally around the fourteenth day of existence. The development of the primitive streak is taken, by such bioethicists, to signify the creation of a unique, human being.[23] In some countries, it is illegal to develop a human embryo for more than 14 days outside a woman's body.[24]

18.5 Additional images

- Human embryo—length, 2 mm. Dorsal view, with the amnion laid open. X 30.

- Lateral section through the mammalian blastodisc.

18.6 References

[1] Mikawa T, Poh AM, Kelly KA, Ishii Y, Reese DE. (2004). "Induction and patterning of the primitive streak, an organizing center of gastrulation in the amniote.". *Dev Dyn* **229** (3): 422–32. doi:10.1002/dvdy.10458. PMID 14991697.

[2] Downs KM. (2009). "The enigmatic primitive streak: prevailing notions and challenges concerning the body axis of mammals". *BioEssays* **31** (8): 892–902. doi:10.1002/bies.200900038. PMC 2949267. PMID 19609969.

[3] Chuai M, Zeng W, Yang X, Boychenko V, Glazier JA, Weijer CJ. (2006). "Cell movement during chick primitive streak formation". *Dev Biol.* 296(1)) (1): 137–49. doi:10.1016/j.ydbio.2006.04.451. PMC 2556955. PMID 16725136.

[4] Chuai M, Weijer CJ. (2008). "The mechanisms underlying primitive streak formation in the chick embryo". *Curr Top Dev Biol.* Current Topics in Developmental Biology **81**: 135–56. doi:10.1016/S0070-2153(07)81004-0. ISBN 978-0-12-374253-7. PMID 18023726.

[5] Khaner O, Eyal-Giladi H. (1989). "The chick's marginal zone and primitive streak formation. I. Coordinative effect of induction and inhibition". *Dev Biol.* **134** (1): 206–14. doi:10.1016/0012-1606(89)90090-0. PMID 2731648.

[6] Perea-Gomez A, Vella FD, Shawlot W, Oulad-Abdelghani M, Chazaud C, Meno C, Pfister V, Chen L, Robertson E, Hamada H, Behringer RR, Ang SL. (2002). "Nodal antagonists in the anterior visceral endoderm prevent the formation of multiple primitive streaks". *Dev Cell.* **3** (5): 745–56. doi:10.1016/S1534-5807(02)00321-0. PMID 12431380.

[7] Cui C, Yang X, Chuai M, Glazier JA, Weijer CJ. (2005). "Analysis of tissue flow patterns during primitive streak formation in the chick embryo". *Dev Biol.* **284** (1): 37–47. doi:10.1016/j.ydbio.2005.04.021. PMID 15950214.

[8] Hatada Y, Stern CD. (1994). "A fate map of the epiblast of the early chick embryo". *Development.* **120** (10): 2879–89. PMID 7607078.

[9] Shah SB, Skromne I, Hume CR, Kessler DS, Lee KJ, Stern CD, Dodd J. (1997). "Misexpression of chick Vg1 in the marginal zone induces primitive streak formation". *Development.* **124** (24): 5127–38. PMID 9362470.

[10] Bachvarova RF, Skromne I, Stern CD. (1998). "Induction of primitive streak and Hensen's node by the posterior marginal zone in the early chick embryo". *Development.* **125** (17): 3521–34. PMID 9693154.

[11] Skromne I, Stern CD. (2001). "Interactions between Wnt and Vg1 signalling pathways initiate primitive streak formation in the chick embryo". *Development.* **128** (15): 2915–27. PMID 11532915.

[12] Skromne I, Stern CD (2002). "A hierarchy of gene expression accompanying induction of the primitive streak by Vg1 in the chick embryo". *Mech Dev.* **114** (1–2): 115–8. doi:10.1016/S0925-4773(02)00034-5. PMID 12175495.

[13] Bertocchini F, Stern CD. (2002). "The hypoblast of the chick embryo positions the primitive streak by antagonizing nodal signaling". *Dev Cell.* **3** (5): 735–44. doi:10.1016/S1534-5807(02)00318-0. PMID 12431379.

[14] Liu P, Wakamiya M, Shea MJ, Albrecht U, Behringer RR, Bradley A (1999). "Requirement for Wnt3 in vertebrate axis formation". *Nat. Genet.* **22** (4): 361–5. doi:10.1038/11932. PMID 10431240.

[15] Zeng L, Fagotto F, Zhang T, Hsu W, Vasicek TJ, Perry WL 3rd, Lee JJ, Tilghman SM, Gumbiner BM, Costantini F. (1997). "The mouse Fused locus encodes Axin, an inhibitor of the Wnt signaling pathway that regulates embryonic axis formation". *Cell.* **90** (1): 181–92. doi:10.1016/S0092-8674(00)80324-4. PMID 9230313.

[16] Pöpperl H, Schmidt C, Wilson V, Hume CR, Dodd J, Krumlauf R, Beddington RS. (1997). "Misexpression of Cwnt8C in the mouse induces an ectopic embryonic axis and causes a truncation of the anterior neuroectoderm". *Development.* **124** (15): 2997–3005. PMID 9247341.

[17] SPRATT NT Jr, HAAS H. (1960). "Integrative mechanisms in development of the early chick blastoderm. I. REgulative potentiality of separated parts". *J Exp Zool.* **145**: 97–137. doi:10.1002/jez.1401450202.

[18] Bertocchini F, Skromne I, Wolpert L, Stern CD. (2004). "Determination of embryonic polarity in a regulative system: evidence for endogenous inhibitors acting sequentially during primitive streak formation in the chick embryo". *Development.* **131** (14): 3381–90. doi:10.1242/dev.01178. PMID 15226255.

[19] Conlon FL, Lyons KM, Takaesu N, Barth KS, Kispert A, Herrmann B, Robertson EJ. (1994). "A primary requirement for nodal in the formation and maintenance of the primitive streak in the mouse". *Development* **120** (7): 1919–28. PMID 7924997.

[20] C Perea-Gomez A, Rhinn M, Ang SL. (2001). "Role of the anterior visceral endoderm in restricting posterior signals in the mouse embryo". *Int J Dev Biol* **45** (1): 311–20. PMID 11291861.

[21] Faure S, de Santa Barbara P, Roberts DJ, Whitman M.. (2002). "Endogenous patterns of BMP signaling during early chick development". *Dev Biol.* **244** (1): 44–65. doi:10.1006/dbio.2002.0579. PMID 11900458.

[22] Streit A, Lee KJ, Woo I, Roberts C, Jessell TM, Stern CD. (1998). "Chordin regulates primitive streak development and the stability of induced neural cells, but is not sufficient for neural induction in the chick embryo". *Development.* **125** (3): 507–19. PMID 9425145.

[23] "The President's Council on Bioethics, Human Cloning and Human Dignity: An Ethical Inquiry. Chapter 6". July 2002.

[24] "Prohibition of Human Cloning for Reproduction Act 2002". Government of Australia Department of Health and Ageing. 22 Dec 2008.

Chapter 19

Primitive pit

The **primitive pit** is a depression in the center of the primitive knot, connecting to the notochord. It consists of part of the primitive streak.

19.1 References

This article incorporates text in the public domain from the 20th edition of Gray's Anatomy (1918)

19.2 External links

- http://www.embryology.ch/anglais/hdisqueembry/triderm01.html

- http://isc.temple.edu/marino/embryology/EMBII97/img020.GIF

- http://cancerweb.ncl.ac.uk/cgi-bin/omd?primitive+pit

- http://embryology.med.unsw.edu.au/Medicine/BGDlab3_11.htm

- http://www.ana.ed.ac.uk/database/humat/notes/embryo/ectoderm/primstr.htm

Chapter 20

Primitive knot

The **primitive knot** (or **primitive node**) is the organizer for gastrulation in vertebrates.

20.1 Diversity

- In birds, it is known as "**Hensen's node**", and is named after its discoverer Victor Hensen.

- In amphibians, it is known as "**Spemann's organizer**", and is named after Hans Spemann (who, with Mangold, first identified the organizer in 1924.[1])

20.2 Development

In chick development, the primitive knot starts as a regional knot of cells that forms on the blastodisc immediately anterior to where the outer layer of cells will begin to migrate inwards - an area known as the primitive streak, which is involved with Koller's sickle. Posterior to the node is the primitive pit, where the cells of the epiblast (the upper layer of embryonic cells) initially begin to invaginate. This invagination expands posteriorly into the primitive groove as the cells layers continue to move into the space between the embryonic cells and the yolk. This differentiates the embryo into the three germ layers - endoderm, mesoderm, and ectoderm. The primitive knot migrates posteriorly as gastrulation proceeds, eventually being absorbed into the tail bud.

20.3 Default model

The cells of the primitive node secrete many cellular signals essential for neural differentiation. After gastrulation the developing embryo is divided into ectoderm, mesoderm, and endoderm. The ectoderm gives rise to epithelial and neural tissue, with neural tissue being the default cell fate.

Bone morphogenetic proteins (BMPs) suppress neural differentiation and promote epithelial growth. Therefore, the primitive node (the dorsal lip of the blastopore) secretes BMP antagonists, including noggin, chordin, and follistatin.

20.4 References

[1] Garcia-Fernàndez J, D'Aniello S, Escrivà H (2007). "Organizing chordates with an organizer". *BioEssays* **29** (7): 619–24. doi:10.1002/bies.20596. PMID 17563072.

20.5 External links

- Overview at nature.com

- Overview at Northwestern University

- Embryonic Organizers at the US National Library of Medicine Medical Subject Headings (MeSH)

Chapter 21

Gastrulation

Gastrulation is a phase early in the embryonic development of most animals, during which the single-layered blastula is reorganized into a trilaminar ("three-layered") structure known as the **gastrula**. These three *germ layers* are known as the ectoderm, mesoderm, and endoderm.[1][2]

Gastrulation takes place after cleavage and the formation of the blastula. Gastrulation is followed by organogenesis, when individual organs develop within the newly formed germ layers.[3] Each layer gives rise to specific tissues and organs in the developing embryo. The **ectoderm** gives rise to epidermis, and to the neural crest and other tissues that will later form the nervous system. The **mesoderm** is found between the ectoderm and the endoderm and gives rise to somites, which form muscle; the cartilage of the ribs and vertebrae; the dermis, the notochord, blood and blood vessels, bone, and connective tissue. The **endoderm** gives rise to the epithelium of the digestive system and respiratory system, and organs associated with the digestive system, such as the liver and pancreas.[4] Following gastrulation, cells in the body are either organized into sheets of connected cells (as in epithelia), or as a mesh of isolated cells, such as mesenchyme.[2][5]

The molecular mechanism and timing of gastrulation is different in different organisms. However, some common features of gastrulation across triploblastic organisms include: (1) A change in the topological structure of the embryo, from a simply connected surface (sphere-like), to a non-simply connected surface (torus-like); (2) the differentiation of cells into one of three types (endodermal, mesodermal, and ectodermal); and (3) the digestive function of a large number of endodermal cells.[6]

Lewis Wolpert, pioneering developmental biologist in the field, has been credited for noting that "It is not birth, marriage, or death, but gastrulation, which is truly the most important time in your life."

The terms "gastrula" and "gastrulation" were coined by Ernst Haeckel, in his 1872 work *"Biology of Calcareous Sponges"*.[7]

Although gastrulation patterns exhibit enormous variation throughout the animal kingdom, they are unified by the five basic types of cell movements that occur during gastrulation: 1) invagination 2) involution 3) ingression 4) delamination 5) epiboly.[8]

21.1 In amniotes

21.1.1 Overview

Gastrulation involves the creation of the **blastopore**, an opening into the archenteron. Note that the blastopore is not an opening into the blastocoel, the space within the blastula, but represents a new inpocketing that pushes the existing surfaces of the blastula together. In amniotes, gastrulation occurs in the following sequence: (1) the embryo becomes asymmetric; (2) the primitive streak forms; (3) cells from the epiblast at the primitive streak undergo an epithelial to mesenchymal transition and ingress at the primitive streak to form the germ layers.[4]

The distinction between protostomes and deuterostomes is based on the direction in which the mouth (stoma) develops in relation to the blastopore. Protostome derives from the Greek word protostoma meaning "first mouth"(πρῶτος + στόμα) whereas Deuterostome's etymology is "second mouth" from the words second and mouth (δεύτερος + στόμα).

The major distinctions between deuterostomes and protostomes are found in embryonic development:

- Mouth/anus

  - In protostome development, the first opening in development, the blastopore, becomes the animal's mouth.

  - In deuterostome development, the blastopore becomes the animal's anus.

- Cleavage

- Protostomes have what is known as *spiral cleavage* which is *determinate*, this meaning that the fate of the cells is determined as they are formed.

- Deuterostomes have what is known as *radial cleavage* that is *indeterminate*.

21.1.2 Loss of symmetry

In preparation for gastrulation, the embryo must become asymmetric along both the proximal-distal axis and the anterior-posterior axis. The proximal-distal axis is formed when the cells of the embryo form the "egg cylinder," which consists of the extraembryonic tissues, which give rise to structures like the placenta, at the proximal end and the epiblast at the distal end. Many signaling pathways contribute to this reorganization, including BMP, FGF, nodal, and Wnt. Visceral endoderm surrounds the epiblast. The distal visceral endoderm (DVE) migrates to the anterior portion of the embryo, forming the "anterior visceral endoderm" (AVE). This breaks anterior-posterior symmetry and is regulated by nodal signaling.[4]

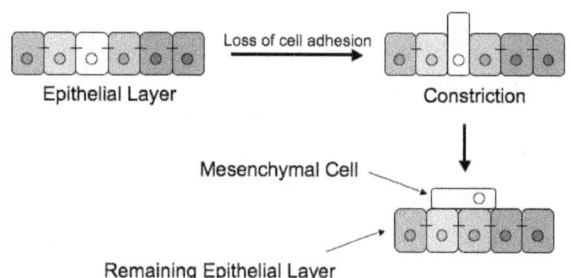

Epithelial to Mesenchmyal Cell Transition – loss of cell adhesion leads to constriction and extrusion of newly mesenchymal cell.

21.1.3 Formation of the primitive streak

The primitive streak is formed at the beginning of gastrulation and is found at the junction between the extraembryonic tissue and the epiblast on the posterior side of the embryo and the site of ingression.[9] Formation of the primitive streak is reliant upon nodal signaling[4] in the Koller's sickle within the cells contributing to the primitive streak and BMP4 signaling from the extraembryonic tissue.[9][10] Furthermore, Cer1 and Lefty1 restrict the primitive streak to the appropriate location by antagonizing nodal signaling.[11] The region defined as the primitive streak continues to grow towards the distal tip.[4]

During the early stages of development, the primitive streak is the structure that will establish bilateral symme-

try, determine the site of gastrulation and initiate germ layer formation. To form the streak, reptiles, birds and mammals arrange mesenchymal cells along the prospective midline, establishing the first embryonic axis, as well as the place where cells will ingress and migrate during the process of gastrulation and germ layer formation.[12] The primitive streak extends through this midline and creates the antero-posterior body axis,[13] becoming the first symmetry-breaking event in the embryo, and marks the beginning of gastrulation.[14] This process involves the ingression of mesoderm and endoderm progenitors and their migration to their ultimate position,[13][15] where they will differentiate into the three germ layers.[12] The localization of the cell adhesion and signaling molecule beta-catenin is critical to the proper formation of the organizer region that is responsible for initiating gastrulation.

21.1.4 Epithelial to mesenchymal transition and ingression

In order for the cells to move from the epithelium of the epiblast through the primitive streak to form a new layer, the cells must undergo an epithelial to mesenchymal transition (EMT) to lose their epithelial characteristics, such as cell-cell adhesion. FGF signaling is necessary for proper EMT. FGFR1 is needed for the up regulation of Snail1, which down regulates E-cadherin, causing a loss of cell adhesion. Following the EMT, the cells ingress through the primitive streak and spread out to form a new layer of cells or join existing layers. FGF8 is implicated in the process of this dispersal from the primitive streak.[11]

21.2 See also

- Blastocyst

- Deuterostome

- Fate mapping

- Hensen's Node[16]

- Invagination

- Neurulation

- Protostome

- Vegetal rotation

21.3 References

21.3.1 Notes

[1] Mundlos 2009: p. 422

[2] McGeady, 2004: p. 34

[3] Hall, 1998: pp. 132-134

[4] Arnold & Robinson, 2009

[5] Hall, 1998: p. 177

[6] Harrison 2011: p. 206

[7] Ereskovsky 2010: p. 236

[8] Gilbert 2010: p. 164.

[9] Tam & Behringer, 1997

[10] Catala, 2005: p. 1535

[11] Tam, P.P. & Loebel, D.A (2007). "Gene function in mouse embryogenesis: get set for gastrulation". *Nat Rev Genet* **8** (5): 368–81. doi:10.1038/nrg2084. PMID 17387317.

[12] Mikawa T, Poh AM, Kelly KA, Ishii Y, Reese DE. (2004). "Induction and patterning of the primitive streak, an organizing center of gastrulation in the amniote.". *Dev Dyn* **229** (3): 422–32. doi:10.1002/dvdy.10458. PMID 14991697.

[13] Downs KM. (2009). "The enigmatic primitive streak: prevailing notions and challenges concerning the body axis of mammals.". *BioEssays* **31** (8): 892–902. doi:10.1002/bies.200900038. PMC 2949267. PMID 19609969.

[14] Chuai M, Zeng W, Yang X, Boychenko V, Glazier JA, Weijer CJ. (2006). "Cell movement during chick primitive streak formation.". *Dev Biol.* 296(1)) (1): 137–49. doi:10.1016/j.ydbio.2006.04.451. PMC 2556955. PMID 16725136.

[15] Chuai M, Weijer CJ. (2008). "The mechanisms underlying primitive streak formation in the chick embryo.". *Curr Top Dev Biol.* **81**: 135–56. doi:10.1016/S0070-2153(07)81004-0. PMID 18023726.

[16] See

21.3.2 Bibliography

- Arnold, Sebastian J.; Robertson, Elizabeth J. (2009). "Making a commitment: cell lineage allocation and axis patterning in the early mouse embryo". *Nat. Rev. Mol. Cell Biol.* **10** (2): 91–103. doi:10.1038/nrm2618. PMID 19129791.

- Catala, Martin (2005). "Embryology of the Spine and Spinal Cord". In Tortori-Donati, Paolo *et al.. Pediatric Neuroradiology: Brain*. Springer. ISBN 978-3-540-41077-5.

- Ereskovsky, Alexander V. (2010). *The Comparative Embryology of Sponges*. Springer. ISBN 978-90-481-8574-0.

- Gilbert, Scott F. (2010). *Developmental Biology* (Ninth ed.). Sinauer Associates. ISBN 978-0-87893-558-1.

- Hall, Brian Keith (1998). "8.3.3 The gastrula and gastrulation". *Evolutionary developmental biology* (2nd ed.). The Netherlands: Kluwer Academic Publishers. ISBN 978-0-412-78580-1.

- Harrison, Lionel G. (2011). *The Shaping of Life: The Generation of Biological Pattern*. Cambridge University Press. ISBN 978-0-521-55350-6.

- McGeady, Thomas A., ed. (2006). "Gastrulation". *Veterinary embryology*. Wiley-Blackwell. ISBN 978-1-4051-1147-8.

- Mundlos, Stefan (2009). "Gene action: developmental genetics". In Speicher, Michael *et al.. Vogel and Motulsky's Human Genetics: Problems and Approaches* (4th ed.). Springer. doi:10.1007/978-3-540-37654-5. ISBN 978-3-540-37653-8.

- Tam, Patrick P.L. & Behringer, Richard R. (1997). "Mouse gastrulation: the formation of a mammalian body plan". *Mech. Dev.* **68** (1-2): 3–25. doi:10.1016/S0925-4773(97)00123-8. PMID 9431800.

21.4 Further reading

- Baron, Margaret H. (2001). "Embryonic Induction of Mammalian Hematopoiesis and Vasculogenesis". In Zon, Leonard I. *Hematopoiesis: a developmental approach*. Oxford University Press. ISBN 978-0-19-512450-7.

- Cullen, K.E. (2009). "embryology and early animal development". *Encyclopedia of life science, Volume 2*. Infobase. ISBN 978-0-8160-7008-4.

- Forgács, G. & Newman, Stuart A. (2005). "Cleavage and blastula formation". *Biological physics of the developing embryo*. Cambridge University Press. ISBN 978-0-521-78337-8.

- Forgács, G. & Newman, Stuart A. (2005). "Epithelial morphogenesis: gastrulation and neurulation". *Biological physics of the developing embryo*. Cambridge University Press. ISBN 978-0-521-78337-8.

- Hart, Nathan H. & Fluck, Richard A. (1995). "Epiboly and Gastrulation". In Capco, David. *Cytoskeletal mechanisms during animal development*. Academic Press. ISBN 978-0-12-153131-7.

- Knust, Elizabeth (1999). "Gastrulation movements". In Birchmeier, Walter & Birchmeier, Carmen. *Epithelial Morphogenesis in Development and Disease*. CRC Press. pp. 152–153. ISBN 978-90-5702-419-1.

- Kunz, Yvette W. (2004). "Gastrulation". *Developmental biology of Teleost fishes*. Springer. ISBN 978-1-4020-2996-7.

- Nation, James L., ed. (2009). "Gastrulation". *Insect physiology and biochemistry*. CRC Press. ISBN 978-0-8493-1181-9.

- Ross, Lawrence M. & Lamperti, Edward D., eds. (2006). "Human Ontogeny: Gastrulation, Neurulation, and Somite Formation". *Atlas of anatomy: general anatomy and musculoskeletal system*. Thieme. ISBN 978-3-13-142081-7.

- Sanes, Dan H. et al. (2006). "Early embryology of metazoans". *Development of the nervous system* (2nd ed.). Academic Press. pp. 1–2. ISBN 978-0-12-618621-5.

- Stanger, Ben Z. & Melton, Douglas A. (2004). "Development of Endodermal Derivatives in the Lungs, Liver, Pancreas, and Gut". In Epstein, Charles J. et al. *Inborn errors of development: the molecular basis of clinical disorders of morphogenesis*. Oxford University Press. ISBN 978-0-19-514502-1.

21.5   External links

- Gastrulation animations

- Gastrulation illustrations and movies from Gastrulation: From Cells To Embryo edited by Claudio Stern

Chapter 22

Primitive groove

A shallow groove, the **primitive groove**, appears on the surface of the primitive streak, and the anterior end of this groove communicates by means of an aperture, the blastophore, with the yolk-sac.

22.1   References

This article incorporates text in the public domain from the 20th edition of Gray's Anatomy (1918)

22.2   External links

- Swiss embryology (from UL, UB, and UF) *hdisqueembry/triderm02*

Chapter 23

Regional differentiation

In the field of developmental biology, **regional differentiation** is the process by which different areas are identified in the development of the early embryo.[1] The process by which the cells become specified differs between organisms.

23.1 Cell fate determination

Main article: Cell fate determination

In terms of developmental commitment, a cell can either be specified or it can be determined. A cell that is specified can have its commitment reversed while the determined state is irreversible.[2] There are two main types of specification: autonomous and conditional. A cell specified autonomously will develop into a specific fate based upon **cytoplasmic determinants** with no regard to the environment the cell is in. A cell specified conditionally will develop into a specific fate based upon other surrounding cells or morphogen gradients.

Specification in sea urchins uses both autonomous and conditional mechanisms to determine the anterior/posterior axis. The anterior/posterior axis lies along the animal/vegetal axis set up during cleavage. The micromeres induce the nearby tissue to become endoderm while the animal cells are specified to become ectoderm. The animal cells are not determined because the micromeres can induce the animal cells to also take on mesodermal and endodermal fates. It was observed that **β-catenin** was present in the nuclei at the vegetal pole of the blastula. Through a series of experiments, one study confirmed the role of β-catenin in the cell-autonomous specification of vegetal cell fates and the micromeres inducing ability.[3] Treatments of LiCl sufficient to vegetalize the embryo resulted in increases in nuclearly localized b-catenin. Reduction of expression of β-catenin in the nucleus correlated with loss of vegetal cell fates. Transplants of micromeres lacking nuclear accumulation of β-catenin were unable to induce a second axis.

For the molecular mechanism of β-catenin and the micromeres, it was observed that Notch was present uniformly on the apical surface of the early blastula but was lost in the secondary mesenchyme cells (SMCs) during late blastula and enriched in the presumptive endodermal cells in late blastula. Notch is both necessary and sufficient for determination of the SMCs. The micromeres express the ligand for Notch, Delta, on their surface to induce the formation of SMCs.

The high nuclear levels of b-catenin results from the high accumulation of the **disheveled protein** at the vegetal pole of the egg. disheveled inactivates GSK-3 and prevents the phosphorylation of β-catenin. This allows β-catenin to escape degradation and enter the nucleus. The only important role of β-catenin is to activate the transcription of the gene **Pmar1**. This gene represses a repressor to allow micromere genes to be expressed.

The Aboral/Oral axis (analogous to the dorsal/ventral axes in other animals) is specified by a Nodal homolog. This Nodal was localized on the future oral side of the embryo. Experiments confirmed that Nodal is both necessary and sufficient to promote development of the oral fate. Nodal also has a role in left/right axis formation.

23.2 Tunicates

Tunicates have been a popular choice for the study of regional specification because tunicates were the first organism in which autonomous specification was discovered and tunicates are evolutionary related to vertebrates.

Early observations in tunicates led to the identification of the **yellow crescent** (also called the myoplasm). This cytoplasm was segregated to future muscle cells and if transplanted could induce the formation of muscle cells. The cytoplasmic determinant **macho-1** was isolated as the necessary and sufficient factor for muscle cell formation. Similar to Sea urchins, the accumulation of b-catenin in the nuclei was identified as both necessary and sufficient to induce endoderm.

Two more cell fates are determined by conditional specification. The endoderm sends a fibroblast growth factor (FGF) signal to specify the notocord and the mesenchyme fates. Anterior cells respond to FGF to become notocord while posterior cells (identified by the presence of macho-1) respond to FGF to become mesenchyme.

The cytoplasm of the egg not only determines cell fate, but also determines the dorsal/ventral axis. The cytoplasm in the vegetal pole specifies this axis and removing this cytoplasm leads to a loss of axis information. The yellow cytoplasm specifies the anterior/posterior axis. When the yellow cytoplasm moves to the posterior of the egg to become posterior vegetal cytoplasm (PVC), the anterior/posterior axis is specified. Removal of the PVC leads to a loss of the axis while transplantation to the anterior reverses the axis.

23.3 *C. elegans*

In the two cell stage, the embryo of the nematode *C. elegans* exhibits mosaic behavior. There are two cells, the **P1** cell and the **AB** cell. The P1 cell was able make all of its fated cells while the AB cell could only make a portion of the cells it was fated to produce. Thus, The first division gives the autonomous specification of the two cells, but the AB cells require a conditional mechanism to produce all of its fated cells.

The AB lineage gives rise to neurons, skin, and pharynx. The P1 cell divides into **EMS** and **P2**. The EMS cell divides into **MS** and **E**. The MS lineage gives rise to pharynx, muscle, and neurons. The E lineage gives rise to intestines. The P2 cell divides into **P3** and **C founder cells**. The C founder cells give rise to muscle, skin, and neurons. The P3 cell divides into **P4** and **D founder cells**. The D founder cells give rise to muscle while the P4 lineage gives rise to the germ line.

- Axis specification

  The anterior/posterior axis is specified by the sperm at the posterior side. At the two cell stage, the anterior cell is the AB cell while the posterior cell is the P1 cell. The dorsal/ventral axis of the animal is set by a random position of cells during the four cell stage of the embryo. The dorsal cell is the ABp cell while the ventral cell is the EMS cell.

- Localization of cytoplasmic determinants

  The autonomous specification of C. elegans arises from different cytoplasmic determinants.

PAR proteins are responsible for partitioning these determinants in the early embryo. These proteins are located at the periphery of the zygote and play a role in intracellular signaling. The current model for the function of these proteins is that they cause local changes in the cytoplasm that lead to different protein accumulation in the posterior vs. the anterior. Mex-5 accumulates in the anterior while PIE-1 and P granules (see below) accumulate in the posterior.

- Specification of germ line

  P granules were identified as the cytoplasmic determinants. While uniformly present at fertilization, these granules become localized in the posterior P1 cell prior to the first division. These granules are further localized between each division into P cells (ex. P2, P3) until after the fourth division when they are put into the P4 cells which become the germ line.

- Specification of EMS and P1 cells

  Other proteins that are likely to function as localized cytoplasmic determinants in the P1 lineage include **SKN-1**, **PIE-1** and **PAL-1**.

  SKN-1 is a cytoplasmic determinant that is localized in the P1 cell lineage and determines EMS cell fate. PIE-1 is localized in the P2 cell lineage and is a general repressor of transcription. SKN-1 is repressed in P2 cells and is unable to specify a EMS fate in these cells. The repressive activity of PIE-1 is required to keep the germ line lineage from differentiating.

- Specification of C and D founder cells

  PAL-1 is required to specify the fates of the C and D founder cells (derived from the P2 lineage). PAL-1, however, is present in both EMS and P2. Normally, PAL-1 activity is repressed in EMS by SKN-1 but not repressed in P2. Both C and D founder cells depend on PAL-1 but there is another factor that is required to distinguish C from D.

- Specification of E lineage

  The specification of the E lineage depends on signals from P2 to the EMS cell. Components of Wnt signaling were involved and were named ***mom*** genes. *mom-2* is a member of the Wnt family of proteins (i.e. the signal) and *mom-5* is a member of the frizzled family of proteins (i.e. the receptor).

- Specification of ABa and ABp

The specification of ABa and ABp depend on another cell-cell signaling event. A difference between these two cell types is that ABa gives rise to anterior pharynx while ABp does not contribute to pharynx. A signal from MS at the 12-cell stage induces pharynx in ABa progeny cells but not in ABp progeny. Signals from the P2 cells prevent the ABp from forming pharynx. This signal from the P2 was discovered to be **APX-1** within the Delta family of proteins. These proteins are known to be ligands for the Notch protein. **GLP-1**, a Notch protein, is also required for specification of the fate of ABp.

23.4 *Drosophila*

See also: Drosophila embryogenesis and Maternal effect

23.4.1 Anterior/posterior axis

The anterior/posterior patterning of *Drosophila* come from three maternal groups of genes. The **anterior group** patterns the head and thoracic segments. The **posterior group** patterns the abdominal segments and the **terminal group** patterns the anterior and posterior terminal regions called the **terminalia** (the **acron** in the anterior and the **telson** in the posterior).

The anterior group genes include **bicoid**. Bicoid functions as a graded morphogen transcription factor that localizes to the nucleus. The head of the embryo forms at the point of highest concentration of bicoid and the anterior pattern depends upon the concentration of bicoid. Bicoid works as a transcriptional activator of the gap genes **hunchback** (hb), buttonhead (btd), empty spiracles (ems), and orthodentical (otd) while also acting to repress translation of **caudal**. A different affinity for bicoid in the promoters of the genes it activates allows for the concentration dependent activation. Otd has a low affinity for bicoid, hb has a higher affinity and so will be activated at a lower bicoid concentration. Two other anterior group genes, **swallow** and **exuperantia** play a role in localizing bicoid to the anterior. Bicoid is directed to the anterior by its 3' untranslated region (3'UTR). The

microtubule cytoskeleton also plays a role in localizing bicoid.

The posterior group genes include **nanos**. Similar to bicoid, nanos is localized to the posterior pole as a graded morphogen. The only role of nanos is to repress the maternally transcribed hunchback mRNA in the posterior. Another protein, **pumilio**, is required for nanos to repress hunchback. Other posterior proteins, oskar (which tethers nanos mRNA), Tudor, vasa, and Valois, localize the germ line determinants and nanos to the posterior.

In contrast to the anterior and the posterior, the positional information for the terminalia come from the follicle cells of the ovary. The terminalia are specified through the action of the **Torso** receptor tyrosine kinase. The follicle cells secrete **Torso-like** into the perivitelline space only at the poles. Torso-like cleaves the pro-peptide **Trunk** which appears to be the Torso ligand. Trunk activates Torso and causes a signal transduction cascade which represses the transcriptional repressor Groucho which in turn causes the activation of the terminal gap genes tailless and huckebein.

23.4.2 Segmentation and homeotic genes

The patterning from the maternal genes work to influence the expression of the **segmentation genes**. The segmentation genes are embryonically expressed genes that specify the numbers, size and polarity of the segments. The **gap genes** are directly influenced by the maternal genes and are expressed in local and overlapping regions along the anterior/posterior axis. These genes are influenced by not only the maternal genes, but also by epistatic interactions between the other gap genes.

The gap genes work to activate the **pair-rule genes**. Each pair-rule gene is expressed in seven stripes as a result of the combined effect of the gap genes and interactions between the other pair-rule genes. The pair-rule genes can be divided into two classes: the **primary pair-rule genes** and the **secondary pair-rule genes**. The primary pair-rules genes are able to influence the secondary pair-rule genes but not vice versa. The molecular mechanism between the regulation of the primary pair-rule genes was understood through a complex analysis of the regulation of even-skipped. Both positive and negative regulatory interactions by both maternal and gap genes and a unique combination of transcription factors work to express even-skipped in different parts of the embryo. The same gap gene can act positively in one stripe but negatively in another.

The expression of the pair-rule genes translate into the expression of the **segment polarity genes** in 14 stripes. The role of the segment polarity genes is to define to boundaries and the polarity of the segments. The means to which the genes accomplish this is believed to involve a wingless and hedgehog graded distribution or cascade of signals initiated by these proteins. Unlike the gap and the pair-rule genes, the segment polarity genes function within cells rather than within the syncytium. Thus, segment polarity genes influence patterning though signaling rather than autonomously. Also, the gap and pair-rule genes are expressed transiently while segment polarity gene expression is maintained throughout development. The continued expression of the segment polarity genes is maintained by a feedback loop involving hedgehog and wingless.

While the segmentation genes can specify the number, size, and polarity of segments, **homeotic genes** can specify the identity of the segment. The homeotic genes are activated by gap genes and pair-rule genes. The **Antennapedia** complex and the **bithorax** complex on the third chromosome contain the major homeotic genes required for specifying segmental identity (actually parasegmental identity). These genes are transcription factors and are expressed in overlapping regions that correlate with their position along the chromosome. These transcription factors regulate other transcription factors, cell surface molecules with roles in cell adhesion, and other cell signals. Later during development, homeotic genes are expressed in the nervous system in a similar anterior/posterior pattern. Homeotic genes are maintained throughout development through the modification of the condensation state of their chromatin. **Polycomb** genes maintain the chromatin in an inactive conformation while **trithorax** genes maintain chromatin in an active conformation.

All homeotic genes share a segment of protein with a similar sequence and structure called the **homeodomain** (the DNA sequence is called the homeobox). This region of the homeotic proteins binds DNA. This domain was found in other developmental regulatory proteins, such as bicoid, as well in other animals including humans. Molecular mapping revealed that the HOX gene cluster has been inherited intact from a common ancestor of flies and mammals which indicates that it is a fundamental developmental regulatory system.

23.4.3 Dorsal/ventral axis

The maternal protein, Dorsal, functions like a graded morphogen to set the ventral side of the embryo (the name comes from mutations which led to a dorsalized phenotype). *Dorsal* is like *bicoid* in that it is a nuclear protein; however, unlike *bicoid*, *dorsal* is uniformly distributed throughout the embryo. The concentration difference arises from differential nuclear transport. The mechanism by which *dorsal* becomes differentially located into the nuclei occurs in three steps.

The first step happens in the dorsal side of the embryo. The nucleus in the oocyte moves along a microtubule track to one side of the oocyte. This side sends a signal, *gurken*, to the *torpedo* receptors on the follicle cells. The *torpedo* receptor is found in all follicle cells; however, the *gurken* signal is only found on the anterior dorsal side of the oocyte. The follicle cells change shape and synthetic properties to distinguish the dorsal side from the ventral side. These dorsal follicle cells are unable to produce the pipe protein required for step two.

The second step is a signal from the ventral follicle cells back to the oocyte. This signal acts after the egg has left the follicle cells so this signal is stored in the perivitelline space. The follicle cells secrete *windbeutel*, *nudel*, and *pipe*, which create a protease-activating complex. Because the dorsal follicle cells do not express *pipe*, they are not able to create this complex. Later, the embryo secretes three inactive proteases (*gastrulation defective*, *snake*, and *Easter*) and an inactive ligand (*spätzle*) into the perivitelline space. These proteases are activated by the complex and cleave *spätzle* into an active form. This active protein is distributed in a ventral to dorsal gradient. *Toll* is a receptor tyrosine kinase for *spätzle* and transduces the graded *spätzle* signal through the cytoplasm to phosphorylate *cactus*. Once phosphorylated, *cactus* no longer binds to *dorsal*, leaving it free to enter the nucleus. The amount of released *dorsal* depends on the amount of *spätzle* protein present.

The third step is the regional expression of zygotic genes *decapentaplegic* (*dpp*), *zerknüllt*, *tolloid*, *twist*, *snail*, and *rhomboid* due to the expression of *dorsal* in the nucleus. High levels of *dorsal* are required to turn on transcription of *twist* and *snail*. Low levels of *dorsal* can activate the transcription of *rhomboid*. *Dorsal* represses the transcription of *zerknüllt*, *tolloid*, and *dpp*. The zygotic genes also interact with each other to restrict their domains of expression.

23.5 Amphibians

23.5.1 Dorsal/ventral axis and organizer

Between fertilization and the first cleavage in *Xenopus* embryos, the cortical cytoplasm of the zygote rotates relative to the central cytoplasm by about 30 degrees to uncover (in some species) a **gray crescent** in the marginal or middle region of the embryo. The **cortical rotation** is powered by microtubules motors moving along parallel arrays of cortical microtubules. This gray crescent marks the future dorsal side of the embryo. Blocking this rotation prevents formation of the dorsal/ventral axis. By the late blastula stage, the *Xenopus* embryos have a clear dorsal/ventral axis.

In the early gastrula, most of the tissue in the embryo is not determined. The one exception is the anterior portion of the dorsal blastopore lip. When this tissue was transplanted to another part of the embryo, it developed as it normally would. In addition, this tissue was able to induce the formation of another dorsal/ventral axis. Hans Spemann named this region the **organizer** and the induction of the dorsal axis the **primary induction**.

The organizer is induced from a dorsal vegetal region called the **Nieuwkoop center**. There are many different developmental potentials throughout the blastula stage embryos. The vegetal cap can give rise to only endodermal cell types while the animal cap can give rise to only epidermal cell types. The marginal zone, however, can give rise to most structures in the embryo including mesoderm. A series of experiments by Pieter Nieuwkoop showed that if the marginal zone is removed and the animal and vegetal caps placed next to each other, the mesoderm comes from the animal cap and the dorsal tissues are always adjacent to the dorsal vegetal cells. Thus, this dorsal vegetal region, named the Nieuwkoop center, was able to induce the formation of the organizer.

Twinning assays identified Wnt proteins as molecules from the Nieuwkoop center that could specify the dorsal/ventral axis. In twinning assays, molecules are injected into the ventral blastomere of a four-cell stage embryo. If the molecules specifies the dorsal axis, dorsal structures will be formed on the ventral side. Wnt proteins were not necessary to specify the axis, but examination of other proteins in the Wnt pathway led to the discovery that **β-catenin** was. β-catenin is present in the nuclei on the dorsal side but not on the ventral side. β-catenin levels are regulated by GSK-3. When active, GSK-3 degrades free β-catenin. There are two possible molecules that might regulate GSK-3: **GBP** (GSK-3 Binding Protein) and **Dishevelled**. The current model is that these act together to inhibit GSK-3 activity. Dishevelled is able to induce a secondary axis when overexpressed and is present at higher levels on the dorsal side after cortical rotation (Symmetry Breaking and Cortical Rotation). Depletion of Dishevelled, however, has no effect. GBP has an effect when depleted and overexpressed. Recent evidence, however, showed that Xwnt11, a Wnt molecule expressed in *Xenopus*, was both sufficient and necessary for dorsal axis formation.[4]

Mesoderm formation comes from two signals: one for the ventral portion and one for the dorsal portion. **Animal cap assays** were used to determine the molecular signals from the vegetal cap that are able to induce the animal cap to form mesoderm. In an animal cap assay, molecules of interest are either applied in medium that the cap is grown in or injected as mRNA in an early embryo. These experiments identified a group of molecules, the **transforming growth factor-β** (TGF-β) family. With dominant negative forms of TGF-β, early experiments were only able to identify the family of molecules involved not the specific member. Recent experiments have identified the ***Xenopus* nodal-related proteins** (Xnr-1, Xnr-2, and Xnr-4) as the mesoderm-inducing signals. Inhibitors of these ligands prevents mesoderm formation and these proteins show a graded distribution along the dorsal/ventral axis.

Vegetally localized mRNA, **VegT** and possibly Vg1, are involved in inducing the endoderm. It is hypothesized that VegT also activates the Xnr-1,2,4 proteins. VegT acts as a transcription factor to activate genes specifying endodermal fate while Vg1 acts as a paracrine factor.

β-catenin in the nucleus activates two transcription factors: **siamois** and **twin**. β-catenin also acts synergistically with VegT to produce high levels of Xnr-1,2,4. Siamois will act synergistically with Xnr-1,2,4 to activate a high level of the transcription factors such as **goosecoid** in the organizer. Areas in the embryo with lower levels of Xnr-1,2,4 will express ventral or lateral mesoderm. Nuclear β-catenin works synergistically with the mesodermal cell fate signal to create the signaling activity of the Nieuwkoop center to induce the formation of the organizer in the dorsal mesoderm.

23.5.2 Organizer function

There are two classes of genes that are responsible for the organizer's activity: transcription factors and secreted proteins. Goosecoid (which has a homology between bicoid and gooseberry) is the first known gene to be expressed in the organizer and is both sufficient and necessary to specify a secondary axis.

The organizer induces ventral mesoderm to become lateral mesoderm, induces the ectoderm to form neural tissue and induces dorsal structures in the endoderm. The mechanism behind these inductions is an inhibition of the **bone morphogenetic protein 4** signaling pathway that ventralizes the embryo. In the absence of these signals, ectoderm reverts to its default state of neural tissue. Four of the secreted molecules from the organizer, **chordin, noggin, follistatin** and ***Xenopus* nodal-related-3** (Xnr-3), directly interact with BMP-4 and block its ability to bind to its receptor. Thus, these molecules create a gradient of BMP-4 along the dorsal/ventral axis of the mesoderm.

BMP-4 mainly acts in trunk and tail region of the embryo while a different set of signals work in the head region. **Xwnt-8** is expressed throughout the ventral and lateral mesoderm. The endomesoderm (can give rise to either endoderm or mesoderm) at the leading edge of the archenteron (future anterior) secrete three factors **Cerberus, Dickkopf,** and **Frzb**. While Cerberus and Frzb bind directly to Xwnt-8 to prevent it from binding to its receptor, Cerberus is also capable of binding to BMP-4 and Xnr1.[5]

Furthermore Dickkopf binds to LRP-5, a transmembrane protein important for the signalling pathway of Xwnt-8, leading to endocytosis of LRP-5 and eventually to an inhibition of the Xwnt-8 pathway.

23.5.3 Anterior/posterior axis

The anterior/posterior patterning of the embryo occurs sometime before or during gastrulation. The first cells to involute have anterior inducing activity while the last cells have posterior inducing activity. The anterior inducing ability comes from the Xwnt-8 antagonizing signals Cereberus, Dickkopf and Frzb discussed above. Anterior head development also requires the function of **IGFs** (insulin-like growth factors) expressed in the dorsal midline and the anterior neural tube. It is believed that IGFs function by activating a signal transduction cascade that interferes and inhibits both Wnt signaling and BMP signaling. In the posterior, two candidates for posteriorizing signals include **eFGF**, a fibroblast growth factor homologue, and retinoic acid.

23.6 Fish

The basis for axis formation in zebrafish parallels what is known in amphibians. The **embryonic shield** has the same function as the dorsal lip of the blastopore and acts as the organizer. When transplanted, it is able to organize a secondary axis and removing it prevents the formation of dorsal structures. β-catenin also has a role similar to its role in amphibians. It accumulates in the nucleus only on the dorsal side; ventral β-catenin induces a secondary axis. It activates the expression of **Squint** (a Nodal related signaling protein aka ndr1) and **Bozozok** (a homeodomain transcription factor similar to Siamois) which act together to activate goosecoid in the embryonic shield.

As in Xenopus, mesoderm induction involves two signals: one from the vegetal pole to induce ventral mesoderm and one from the Nieuwkoop center equivalent dorsal vegetal cells to induce dorsal mesoderm.

The signals from the organizer also parallel to those from amphibians. Noggin and chordin homologue **Chordino**, binds to a BMP family member, BMP2B, to block it from ventralizing the embryo. Dickkopf binds to a Wnt homolog Wnt8 to block it from ventralizing and posteriorizing the embryo.

There is a third pathway regulated by β-catenin in fish. β-catenin activates the transcription factor **stat3**. Stat3 coordinates cell movements during gastrulation and contributes to establishing planar polarity.

23.7 Birds

The dorsal/ventral axis is defined in chick embryos by the orientation of the cells with respect to the yolk. Ventral is down with respect to the yolk while animal is up. This axis is defined by the creation of a pH difference "inside" and "outside" of the blastoderm between the subgerminal space and the albumin on the outside. The subgerminal space has a pH of 6.5 while the albumin on the outside has a pH of 9.5.

The anterior/posterior axis is defined during the initial tilting of the embryo when the eggshell is being desposited. The egg is constantly being rotated in a consistent direction and there is a partial stratification of the yolk; the lighter yolk components will be near one end of the blastoderm and will become the future posterior. The molecular basis of the posterior is not known, however, the accumulation of cells eventually results in the posterior marginal zone (PMZ).

The PMZ is the equivalent of the Nieuwkoop center is that its role is to induce Hensen's node. Transplantation of the PMZ results in induction of a primitive streak, however, PMZ does not contribute to the streak itself. Similar to the Nieuwkoop center, the PMZ expresses both Vg1 and nuclear localized β-catenin.

The Hensen's node is equivalent to the organizer. Transplantation of Hensen's node results in the formation of a secondary axis. Hensen's node is the site where gastrulation begins and it becomes the dorsal mesoderm. Hensen's node is formed from the induction of PMZ on the anterior part of the PMZ called Koller's sickle. When the primitive streak forms, these cells expand out to become Hensen's node. These cells express goosecoid consistent with their role as the organizer.

The function of the organizer in chick embryos is similar to that of amphibians and fish, however, there are some differences. Similar to the amphibians and fish, the organizer does secrete Chordin, Noggin and Nodal proteins that antagonize BMP signaling and dorsalize the embryo. Neural induction, however, does not rely entirely on inhibiting the BMP signaling. Overexpression of BMP antagonists is not enough induce formation of neurons nor overexpressing BMP block formation of neurons. While the whole story is unknown for neural induction, FGFs seem to play a role in mesoderm and neural induction. The anterior/posterior patterning of the embryo requires signals like cereberus from the hyboplast and the spatial regulation of retinoic acid accumulation to activate the 3' Hox genes in the posterior neuroectoderm (hindbrain and spinal cord).

23.8 Mammals

The earliest specification in mouse embryos occurs between trophoblast and inner cell mass cells in the outer polar cells and the inner apolar cells respectively. These two groups become specified at the eight-cell stage during compaction, but do not become determined until they reach the 64-cell stage. If an apolar cell is transplanted to the outside during the 8-32 cell stage, that cell will develop as a trophoblast cell.

The anterior/posterior axis in the mouse embryo is specified by two signaling centers. In the mouse embryo, the egg forms a cylinder with the epiblast forming a cup at the distal end of that cylinder. The epiblast is surrounded by the visceral endoderm, the equivalent of the hypoblast of humans and chicks. Signals for the anterior/posterior axis come from primitive knot. The other important site is the **anterior visceral endoderm** (AVE). The AVE lies anterior to the node's most anterior position and lies just under the epiblast in the region that will become occupied by migrating endomesoderm to form head mesoderm and foregut endoderm. The AVE interacts with the node to specify the most anterior structures. Thus, the node is able to form a normal trunk, but requires signals from the AVE to form a head.

The discovery of the homeobox in *Drosophila* flies and its conservation in other animals has led to advancements in understanding the anterior/posterior patterning. Most of the Hox genes in mammals show an expression pattern that parallels the homeotic genes in flies. In mammals, there are four copies of the Hox genes. Each set of Hox genes are **paralogous** to the others (Hox1a is a paralogue of Hox1b, etc.) These paralogs show overlapping expression patterns and could act redundantly. However, double mutations in paralogous genes can also act synergistically indicating that the genes must work together for function.

23.9 See also

- Specification (technical standard)
- Pattern formation

23.10 References

[1] Slack, J.M.W. (2013) Essential Developmental Biology. Wiley-Blackwell, Oxford.

[2] Slack, J.M.W. (1991) From egg to embryo. Regional specification in early development. Cambridge University Press, Cambridge

[3] McClay D, Peterson R, Range R, Winter-Vann A, Ferkowicz M (2000). "A micromere induction signal is activated by beta-catenin and acts through notch to initiate specification of secondary mesenchyme cells in the sea urchin embryo.". *Development* **127** (23): 5113–22. PMID 11060237.

[4] Tao Q, Yokota C, Puck H, Kofron M, Birsoy B, Yan D, Asashima M, Wylie C, Lin X, Heasman J (2005). "Maternal wnt11 activates the canonical wnt signaling pathway required for axis formation in *Xenopus* embryos.". *Cell* **120** (6): 857–71. doi:10.1016/j.cell.2005.01.013. PMID 15797385.

[5] Silva, A C; Filipe M; Kuerner K M K; Steinbeisser H; BelocJ A (Oct 2003). "Endogenous Cerberus activity is required for anterior head specification in Xenopus.". *Development* (England) **130** (20): 4943–53. doi:10.1242/dev.00705. ISSN 0950-1991. PMID 12952900.

Chapter 24

Embryonic disc

The floor of the amniotic cavity is formed by the **embryonic disc** (or **embryonic disk**) composed of a layer of prismatic cells, the embryonic ectoderm, derived from the inner cell-mass and lying in apposition with the endoderm.

In humans, it is the stage of development that occurs after implantation and prior to the embryonic folding (e.g. seen between about day 14 to day 21 post fertilization). It is derived from the epiblast layer, which lies between the hypoblast layer and the amnion. The epiblast layer is derived from the inner cell mass. Through the process of gastrulation, the bilaminar embryonic disc becomes trilaminar. The notochord forms thereafter. Through the process of neurulation, the notochord induces the formation of the neural tube in the embryonic disc.

24.1 References

This article incorporates text in the public domain from the 20th edition of Gray's Anatomy (1918)

24.2 External links

- Diagram at manchester.ac.uk

Chapter 25

Ectoderm

Ectoderm is one of the three primary germ layers in the very early embryo. The other two layers are the mesoderm (middle layer) and endoderm (most proximal layer), with the ectoderm as the most exterior (or distal) layer.[1] It emerges and originates from the outer layer of germ cells. The word ectoderm comes from the Greek *ektos* meaning "outside", and *derma*, meaning "skin."[2]

Generally speaking, the ectoderm differentiates to form the nervous system (spine, peripheral nerves and brain),[3][4] tooth enamel and the epidermis (the outer part of integument). It also forms the lining of mouth, anus, nostrils, sweat glands, hair and nails.[4]

In vertebrates, the ectoderm has three parts: external ectoderm (also known as surface ectoderm), the neural crest, and neural tube. The latter two are known as neuroectoderm.

25.1 History

Heinz Christian Pander, a Russian biologist, has been credited for the discovery of the three germ layers that form during embryogenesis. Pander received his doctorate in zoology from the University of Wurzburg in 1817. He began his studies in embryology using chicken eggs, which allowed for his discovery of the ectoderm, mesoderm and endoderm. Due to his findings, Pander is sometimes referred to as the "founder of embryology". Pander's work of the early embryo was continued by a Prussian-Estonian biologist named Karl Ernst von Baer. Baer took Pander's concept of the germ layers and through extensive research of many different types of species, he was able to extend this principle to all vertebrates. Baer also received credit for the discovery of the blastula. Baer published his findings, including his germ layer theory, in a textbook which translates to *On the Development of Animals* which he released in 1828.[5]

25.2 Differentiation

25.2.1 Initial appearance

The ectoderm can first be observed in amphibians and fish during the later stages of a process called gastrulation. At the start of this process, the developing embryo has divided into many cells separating the embryo, which is now a hollow sphere of cells called the blastula, into two parts, the animal hemisphere and vegetal hemisphere. It is the animal hemisphere of the blastula that will eventually become the ectoderm.[2]

25.2.2 Early development

Like the other two germ layers, mesoderm and endoderm, the ectoderm forms shortly after the egg is fertilized, and rapid cell division initiates. The epidermis of the skin originates from the less dorsal ectoderm which surrounds the neuroectoderm at the early gastrula stage of embryonic development. The position of the ectoderm relative to the other germ layers of the embryo is governed by "selective affinity", meaning that the inner surface of the ectoderm has a strong (positive) affinity for the mesoderm, and a weak (negative) affinity for the endoderm layer. This selective affinity changes during different stages of development. The strength of the attraction between two surfaces of two germ layers is determined by the amount and type of cadherin molecules present on the cells' surface. For example, the expression of N-cadherin is crucial to maintaining separation of precursor neural cells from precursor epithelial cells.[2] The ectoderm is instructed to become the nervous system by the notochord, which is typically positioned above it.[2]

Gastrulation

During the process of gastrulation, a special type of cells called bottle cells invaginates a hole on the surface of the blastula which is called the dorsal lip of the blastopore. Once this lip has been established, the bottle cells will extend inward and migrate along the inner wall of the blastula known as the roof of the blastocoel. The once superficial cells of the animal pole are destined to become the cells of the middle germ layer called the mesoderm. Through the process of radial extension, cells of the animal pole that were once several layers thick divide to from a thin layer. At the same time, when this thin layer of dividing cells reaches the dorsal lip of the blastopore, another process occurs termed convergent extension. During convergent extension, cells that approach the lip intercalate mediolaterally, in such a way that cells are pulled over the lip and inside the embryo. These two processes allow for the prospective mesoderm cells to be placed between the ectoderm and the endoderm. Once convergent extension and radial intercalation are underway, the rest of the vegetal pole, which will become endoderm cells, is completely engulfed by the prospective ectoderm, as these top cells undergo epiboly, where the ectoderm cells divide in a way to form one layer. This creates a uniform embryo composed of the three germ layers in their respective positions.[2]

25.2.3 Later development

Once there is an embryo with three established germ layers, differentiation among these three layers proceeds. The next event that will take place within the ectoderm is the process of neurulation, which results in the formation of the neural tube, neural crest cells and the epidermis. It is these three components of the ectoderm that will each give rise to a particular set of cells. The neural tube cells will become the central nervous system, neural crest cells will become the peripheral and enteric nervous system, along with melanocytes, facial cartilage and the dentin of teeth, and the epidermal cell region will give rise to epidermis, hair, nails, sebaceous glands, olfactory and mouth epithelium, as well as eyes.[2]

Neurulation

Neurulation proceeds by primary and secondary neurulation, both positioning neural crest cells between a superficial epidermal layer and a deep neural tube. During primary neurulation, the notochord cells of the mesoderm signal the adjacent, superficial ectoderm cells to reposition themselves in a columnar pattern to form cells of the ectodermal neural plate.[6] As the cells continue to elongate, a group of cells immediately above the notochord change their shape, forming a wedge in the ectodermal region. These special cells are called medial hinge cells (MHP). Now, as the ectoderm continues to elongate, the ectodermal cells of the neural plate fold inward. The inward folding of the ectoderm by

virtue of mainly cell division continues until another group of cells form within the neural plate. These cells are termed dorsolateral hinge cells (DLHP), and once formed, the inward folding of the ectoderm stops. The DLHP cells function in a similar fashion as MHP cells regarding their wedge like shape, however, the DLHP cells result in the ectoderm converging. This convergence is led by ectodermal cells above the DLHP cells known as the neural crest. The neural crest cells eventually pull the adjacent ectodermal cells together, which leaves neural crest cells between the prospective epidermis and hollow, neural tube.[2]

Organogenesis

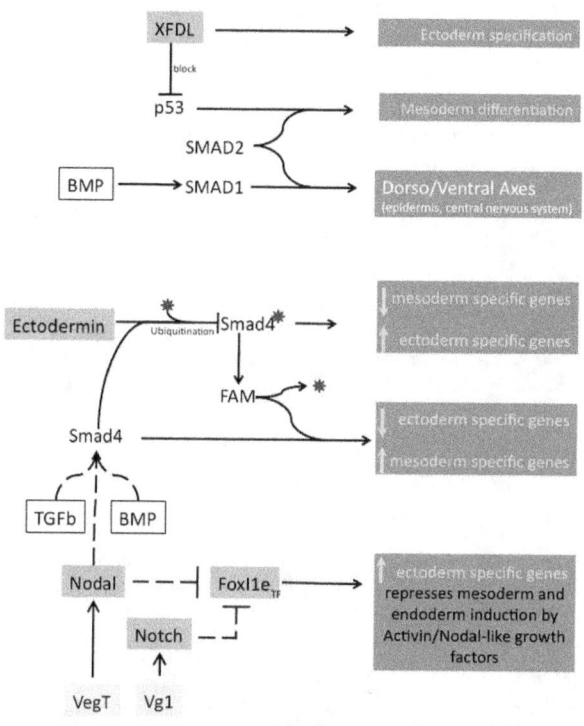

Ectodermal specification

All of the organs that rise from the ectoderm such as the nervous system, teeth, hair and many exocrine glands, originate from two adjacent tissue layers: the epithelium and the mesenchyme. [7] Several signals mediate the organogenesis of the ectoderm such as: FGF, TGFβ, Wnt, and regulators from the hedgehog family. The specific timing and manner that the ectodermal organs form is dependent on the invagination of the epithelial cells.[8] FGF-9 is an important factor during the initiation of tooth germ development. The rate of epithelial invagination in significantly increased by action of FGF-9, which is only expressed in the epithelium, and not in the mesenchyme. FGF-10 helps to stimulate epithelial cell proliferation, in order make larger tooth germs. Mammalian teeth develop from ectoderm derived from the mesenchyme: oral ectoderm and neural crest. The epithelial components of the stem cells for continuously growing teeth form from tissue layers called the stellate reticulum and the suprabasal layer of the surface ectoderm.[8]

25.3   Clinical significance

25.3.1   Ectodermal dysplasia

Ectodermal dysplasia is a rare but severe condition where the tissue groups (specifically teeth, skin, hair, nails and sweat glands) derived from the ectoderm undergo abnormal development. Ectodermal dysplasia is a vague term, as there are over 170 subtypes of ectodermal dysplasia. It has been accepted that the disease is caused by a mutation or a combination of mutations in a number of genes. Research of the disease is ongoing, as only a fraction of the mutations involved with an ectodermal dysplasia subtype have been identified.[9]

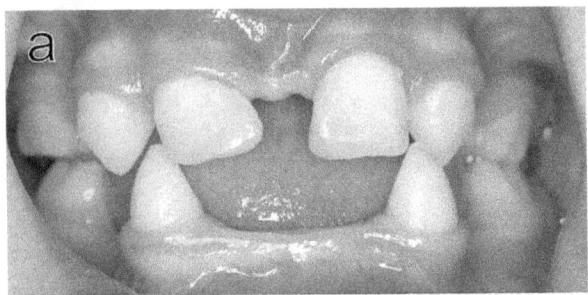

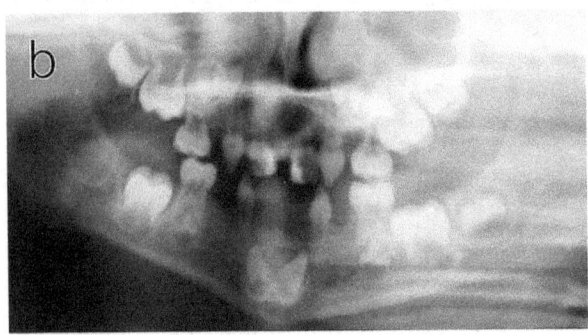

Dental abnormalities in a 5-year-old girl from north Sweden family who suffered from various symptoms of autosomal dominant hypohidrotic ectodermal dysplasia (HED) a) Intraoral view. Note that the upper incisors have been restored with composite material to disguise their original conical shape. b) Orthopantomogram showing absence of ten primary and eleven permanent teeth in the jaws of the same individual.

Hypohidrotic ectodermal dysplasia (HED) is the most common subtype of the disease. Clinical cases of patients with this condition displayed a range of symptoms. One of the common abnormalities of HED is hypohidrosis, or the inability to sweat, which can be attributed to dysfunctional

sweat glands. This aspect can be especially dangerous in warm climates where the patient could potentially suffer from hyperthermia. Facial malformations are also related to HED such as disfigured or absent teeth, wrinkled skin around the eyes, misshaped nose along with scarce and thin hair. Skin problems, like eczema have also been observed in cases.[10] It typically follows an X-linked recessive pattern of inheritance of the EDA genes.[11] This disease typically affects males because they have only one X chromosome, meaning only one copy of the mutated gene is enough to cause abnormal development. For females to be affected, both X chromosomes would need to carry the gene mutation. If a female has a mutated version of the gene on one X chromosome, they are considered carrier of the disease.

25.4   See also

- Ectoderm specification

- Coelom

- Embryology

- Endoderm

- Gastrulation

- Mesoderm

- Neural plate

25.5   References

[1] Langman's Medical Embryology, 11th edition. 2010.

[2] Gilbert, Scott F. Developmental Biology. 9th ed. Sunderland, MA: Sinauer Associates, 2010: 333-370. Print.

[3] http://www.bioethics.gov/reports/stemcell/glossary.html

[4] http://simple.wikipedia.7val.com/wiki/Mate

[5] Baer KE von (1986) In: Oppenheimer J (ed.) and Schneider H (transl.), Autobiography of Dr. Karl Ernst von Baer. Canton, MA: Science History Publications.

[6] O'Rahilly, R; Müller, F (1994). "Neurulation in the normal human embryo". *Ciba Found Symp.* **181**: 70–82. PMID 8005032.

[7] Pispa, J; Thesleff, I (Oct 15, 2003). "Mechanisms of ectodermal organogenesis.". *Developmental Biology* **262** (2): 195–205. doi:10.1016/S0012-1606(03)00325-7. PMID 14550785.

[8] Tai, Y. Y.; Chen, R. S.; Lin, Y.; Ling, T. Y.; Chen, M. H. (2012). "FGF-9 accelerates epithelial invagination for ectodermal organogenesis in real time bioengineered organ manipulation". *Cell Communication and Signaling* **10** (1): 34. doi:10.1186/1478-811X-10-34. PMC 3515343. PMID 23176204.

[9] Priolo, M.; Laganà, C (September 2001). "Ectodermal Dysplasias: A New Clinical-Genetic Classification". *Journal of Medical Genetics* **38** (9): 579–585. doi:10.1136/jmg.38.9.579. PMID 11546825.

[10] Clarke, A., D. I. Phillips, R. Brown, and P. S. Harper. "Clinical Aspects of X-linked Hypohidrotic Ectodermal Dysplasia." Archives of Disease in Childhood 62.10 (1987): 989-96. Print.

[11] Bayes, M.; Hartung, A. J.; Ezer, S.; Pispa, J.; Thesleff, I.; Srivastava, A. K.; Kere, J. (1998). "The Anhidrotic Ectodermal Dysplasia Gene (EDA) Undergoes Alternative Splicing and Encodes Ectodysplasin-A with Deletion Mutations in Collagenous Repeats". *Human Molecular Genetics* **7** (11): 1661–1669. doi:10.1093/hmg/7.11.1661. PMID 9736768.

Chapter 26

Surface ectoderm

The **surface ectoderm** (or **external ectoderm**) forms the following structures:

- Skin (only epidermis; dermis is derived from mesoderm) (along with glands, hair, and nails)

- Epithelium of the mouth and nasal cavity salivary glands, and glands of mouth and nasal cavity

- Tooth enamel (as a side note, dentin and dental pulp are formed from ectomesenchyme which is derived from ectoderm (specifically neural crest cells and travels with mesenchmyal cells)

- Epithelium of anterior pituitary

- Lens, cornea, lacrimal gland, tarsal glands and the conjunctiva of the eye

- Apical ectodermal ridge inducing development of the limb buds of the embryo.

- Sensory receptors in epidermis

26.1 References

This article incorporates text in the public domain from the 20th edition of Gray's Anatomy (1918)

26.2 External links

- http://cwx.prenhall.com/bookbind/pubbooks/ martini10/chapter18/custom3/deluxe-content.html

- Thomas, Jane Coad with Melvyn Dunstall ; foreword by Meryl (2001). *Anatomy and physiology for midwives*. Edinburgh ; New York: Mosby. ISBN 0723429790.

Chapter 27

Neuroectoderm

Neuroectoderm (or **neural ectoderm** or **neural tube epithelium**) is ectoderm which receives bone morphogenetic protein-inhibiting signals from proteins such as noggin, which leads to the development of the nervous system from this tissue.

After recruitment from the ectoderm, the neuroectoderm undergoes three stages of development: transformation into the neural plate, transformation into the neural groove (with associated neural folds), and transformation into the neural tube. After formation of the tube, the brain forms into three sections; the hindbrain, the midbrain, and the forebrain.

The types of neuroectoderm include:

- Neural crest
  - pigment cells in the skin
  - ganglia of the autonomic nervous system
  - dorsal root ganglia.
  - facial cartilage
  - aorticopulmonary septum of the developing heart and lungs
  - ciliary body of the eye
  - adrenal medulla
  - parafollicular cells in the thyroid
- Neural tube
  - brain (rhombencephalon, mesencephalon and prosencephalon)
  - spinal cord and motor neurons
  - retina
  - posterior pituitary

27.1 See also

- neural plate
- neuroectodermal tumor
- neuroepithelial cell

27.2 References

This article incorporates text in the public domain from the 20th edition of Gray's Anatomy (1918)

27.3 External links

- *bdyfm-007*—Embryo Images at University of North Carolina
- http://sprojects.mmi.mcgill.ca/embryology/earlydev/week3/neurulation.html
- http://www.med.umich.edu/lrc/coursepages/M1/embryology/embryo/08nervoussystem.htm

Chapter 28

Somatopleuric mesenchyme

In the anatomy of an embryo, the **somatopleuric mesenchyme** is a structure created during embryogenesis when the lateral mesoderm splits into two layers. The outer (or somatic) layer becomes applied to the inner surface of the ectoderm, and with it forms the somatopleure.

28.1   See also

- Splanchnopleure

28.2   References

This article incorporates text in the public domain from the 20th edition of Gray's Anatomy (1918)

28.3   External links

- Diagram at Yuba Community College District
- Overview at Kennesaw State University

Chapter 29

Neurulation

Neurulation refers to the folding process in vertebrate embryos, which includes the transformation of the neural plate into the neural tube.[1] The embryo at this stage is termed the neurula.

The process begins when the notochord induces the formation of the central nervous system (CNS) by signaling the ectoderm germ layer above it to form the thick and flat neural plate. The neural plate folds in upon itself to form the neural tube, which will later differentiate into the spinal cord and the brain, eventually forming the central nervous system.[2]

Different portions of the neural tube form by two different processes, called primary and secondary neurulation, in different species.

- In **primary neurulation**, the neural plate creases inward until the edges come in contact and fuse.

- In **secondary neurulation**, the tube forms by hollowing out of the interior of a solid precursor.

29.1 Primary neurulation

29.1.1 Induction

Primary neurulation occurs in response to soluble growth factors secreted by the notochord. Ectodermal cells are induced to form neuroectoderm from a variety of signals. Ectoderm sends and receives signals of bone morphogenetic protein 4 (BMP4) and cells which receive BMP4 signal develop into epidermis. The inhibitory signals chordin, noggin and follistatin are needed to form neural plate. These inhibitory signals are created and emitted by the Spemann organiser. Cells which do not receive BMP4 signaling due to the effects of the inhibitory signals will develop into the anterior neuroectoderm cells of the neural plate. Cells which receive fibroblast growth factor (FGF) in addition to the inhibitory signals form posterior neural plate cells.

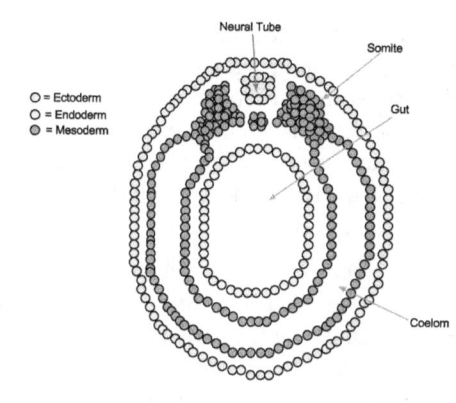

Cross section of a vertebrate embryo in the neurula stage

29.1.2 Shape change

The cells of the neural plate are signaled to become high-columnar and can be identified through microscopy as different from the surrounding epiblastic ectoderm. The cells move laterally and away from the central axis and change into a truncated pyramid shape. This pyramid shape is achieved through tubulin and actin in the apical portion of the cell which constricts as they move. The variation in cell shapes is partially determined by the location of the nucleus within the cell, causing bulging in areas of the cells forcing the height and shape of the cell to change. This process is known as apical constriction

29.1.3 Folding

The process of the flat neural plate folding into the cylindrical neural tube is termed **primary neurulation**. As a result of the cellular shape changes, the neural plate forms the medial hinge point (MHP) . The expanding epidermis puts pressure on the MHP and causes the neural plate to fold resulting in neural folds and the creation of the neural groove. The neural folds form dorsolateral hinge points (DLHP) and pressure on this hinge causes the neural folds to meet and fuse at the midline. The fusion requires the regulation of cell adhesion molecules. The neural plate switches from E-cadherin expression to N-cadherin and N-CAM expression to recognize each other as the same tissue and close the tube. This change in expression stops the binding of the neural tube to the epidermis. Neural plate folding is a complicated step.

The notochord plays an integral role in the development of the neural tube. Prior to neurulation, during the migration of epiblastic endoderm cells towards the hypoblastic endoderm, the notochordal process opens into an arch termed the **notochordal plate** and attaches overlying neuroepithelium of the neural plate. The notochordal plate then serves as an anchor for the neural plate and pushes the two edges of the plate upwards while keeping the middle section anchored. Some of the notochodral cells become incorporated into the center section neural plate to later form the floor plate of the neural tube. The notochord plate separates and forms the solid notochord.

The folding of the neural tube to form an actual tube does not occur all at once. Instead, it begins approximately at the level of the fourth somite at Carnegie stage 9 (around Embryonic day 20 in humans). The lateral edges of the neural plate touch in the midline and join together. This continues both cranially (toward the head) and caudally (toward the tail). The openings that are formed at the cranial and caudal regions are termed the **cranial and caudal neuropores**. In human embryos, the cranial neuropore closes approximately on day 24 and the caudal neuropore on day 28.[3] Failure of the cranial (anterior) and caudal (posterior) neuropore closure results in conditions called anencephaly and spina bifida, respectively. Additionally, failure of the neural tube to close throughout the length of the body results in a condition called rachischisis.[4]

29.1.4   Patterning

After sonic hedgehog (SHH) signalling from the notochord induces its formation, the floor plate of the incipient neural tube also secretes SHH. After closure, the neural tube forms a basal or floor plate and a roof or alar plate in response to the combined effects of SHH and factors including BMP4 secreted by the roof plate. The basal plate forms most of the ventral portion of the nervous system, including the motor portion of the spinal cord and brain stem; the alar plate forms the dorsal portions, devoted mostly to sensory processing.[5]

The dorsal epidermis expresses BMP4 and BMP7. The roof plate of the neural tube responds to those signals to express more BMP4 and other transforming growth factor beta (TGF-β) signals to form a dorsal/ventral gradient

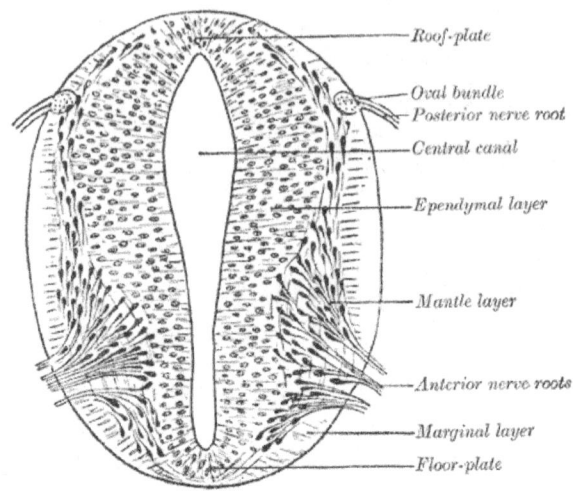

Transverse section of the neural tube showing the floor plate and roof plate

among the neural tube. The notochord expresses SHH. The floor plate responds to SHH by producing its own SHH and forming a gradient. These gradients allows for the differential expression of transcription factors.[6]

29.1.5   Complexities of the model

In actuality, the folding of the neural tube is still not entirely understood and is still being studied. The simplistic model of the closure occurring in one step cranially and caudally does not explain the high frequency of neural tube defects (see below). Proposed theories include that closure of the neural tube occurs in regions, rather than entirely linearly.

29.2   Secondary neurulation

In secondary neurulation, the neural ectoderm and some cells from the endoderm form the medullary cord. The medullary cord condenses, separates and then forms cavities. These cavities then merge to form a single tube. Secondary neurulation occurs in the posterior section of most animals but it is better expressed in birds. Tubes from both primary and secondary neurulation eventually connect.[7]

29.3   Early brain development

The anterior segment of the neural tube forms the three main parts of the brain: the forebrain, midbrain, and the hindbrain. Formation of these structures begins with a swelling of the neural tube in a pattern specified by Hox genes. Ion pumps are used to increase the fluid pressure within the tube and create a bulge. A blockage between the brain and the spinal cord prevents the fluid accumulation from leaking out. These brain regions further divide into subregions. The hindbrain divides into different segments called rhombomeres. Neural crest cells form ganglia above each rhombomere. The neural tube becomes the germinal neuroepithelium and serves as a source of new neurons during brain development. The brain develops from the inside-out.

29.4　Non-neural ectoderm tissue

Paraxial mesoderm surrounding the notochord at the sides will develop into the somites (future muscles, bones, and contributes to the formation of limbs of the vertebrate).

29.5　Neural crest cells

Main article: Neural crest

Masses of tissue called the neural crest that are located at the very edges of the lateral plates of the folding neural tube separate from the neural tube and migrate to become a variety of different but important cells.

Neural crest cells will migrate through the embryo and will give rise to several cell populations, including pigment cells and the cells of the peripheral nervous system.

29.6　Neural tube defects

Closure of the neural tube occurs in the middle, and then moves superiorly and inferiorly. Failure to close the superior neural tube results in anencephaly, a condition characterised by forebrain and skull degeneration, which is always fatal. Failure to close the inferior tube is known as spina bifida, which in its most severe form is characterised by failure to form the neural plate (bifida is derived from Latin, to mean 'cleft in two parts'). Less severe forms are characterised by defects in the meninges and vertebrae which overlie the posterior spinal cord.[8]

Spina bifida can lead to paralysis beneath the affected region of the spinal cord. Sufferers may require crutches or wheelchairs to move about, and may also suffer from lack of bladder and bowel control.

Neural tube defects are among the most common and disabling birth defects, occurring in roughly 1 in every 500 live births.[9]

29.7　See also

- Neural fold
- Neural plate
- Neural crest

29.8　References

[1] Larsen WJ. Human Embryology. Third ed. 2001.P 86. ISBN 0-443-06583-7

[2] "Chapter 14. Gastrulation and Neurulation". biology.kenyon.edu. Retrieved 2 February 2016.

[3] Youman's Neurological Surgery, H Richard Winn, 6th ed. Volume 1, p 81, 2011 Elsevier Saunders, Philadelphia, PA

[4] Gilbert, SF (2000). "12: Formation of the Neural Tube". Developmental Biology (6 ed.). Sunderland, MA: Sinauer Associates. ISBN 978-0-87893-243-6. Retrieved 30 November 2011.

[5] Gilbert, SF (2013). "10: Emergence of the Ectoderm". Developmental Biology (10 ed.). Sunderland, MA: Sinauer Associates. ISBN 978-0-87893-978-7. Retrieved 22 March 2015.

[6] Gilbert, SF (2013). "10: Emergence of the Ectoderm". Developmental Biology (10 ed.). Sunderland, MA: Sinauer Associates. ISBN 978-0-87893-978-7. Retrieved 22 March 2015.

[7] Shimokita, E; Takahashi, Y (April 2011). "Secondary neurulation: Fate-mapping and gene manipulation of the neural tube in tail bud.". Development, growth & differentiation 53 (3): 401–10. doi:10.1111/j.1440-169X.2011.01260.x. PMID 21492152.

[8] Bear, Mark (2009). Neuroscience - Exploring the Brain. LWW. pp. 182–183.

[9] Daley, Darrel. Formation of the Nervous System. Last accessed on Oct 29, 2007.

29.9　Further reading

- Almeida, Karla L.; et al. (2010). "Neural Induction". In Henning, Ulrich. Perspectives of Stem Cells: From Tools for Studying Mechanisms of Neuronal Differentiation Towards Therapy. Springer. ISBN 978-90-481-3374-1.

- Basch, Martín L. & Bonner-Fraser, Marianne (2006). "Neural Crest Inducing Signals". In Saint-Jennet, Jean-Pierre. Neural crest induction and differentiation. Springer. ISBN 978-0-387-35136-0.

- Harland, Richard M. (1997). "Neural induction in *Xenopus*". In Cowan, W. Maxwell. *Molecular and cellular approaches to neural development*. Oxford University Press. ISBN 978-0-19-511166-8.

- Ladher, Raj & Schoenwolf, Gary C. (2004). "Making a neural tube". In Jacobson, Marcus & Rao, Mahendra S. *Developmental neurobiology*. Springer. ISBN 978-0-306-48330-1.

- Tian, Jing & Sampath, Karuna (2004). "Formation and Functions of the Floor Plate". In Gong, Zhiyuan & Korzh, Vladimir. *Fish development and genetics: the zebrafish and medaka models*. World Scientific. pp. 123; 139–140. ISBN 978-981-238-821-6.

- Zhang, Su-Chun (2005). "Neural specification from human embryonic stem cells". In Odorico, John S. et al. *Human embryonic stem cells*. Garland Science. ISBN 978-1-85996-278-7.

29.10 External links

- Overview at uvm.edu

- Neurulation Animation

Chapter 30

Neural crest

Neural crest cells are a temporary group of cells unique to vertebrates that arise from the embryonic ectoderm cell layer, and in turn give rise to a diverse cell lineage -- including melanocytes, craniofacial cartilage and bone, smooth muscle, peripheral and enteric neurons and glia.[1]

After gastrulation, neural crest cells are specified at the border of the neural plate and the non-neural ectoderm. During neurulation, the borders of the neural plate, also known as the neural folds, converge at the dorsal midline to form the neural tube. Subsequently, neural crest cells from the roof plate of the neural tube undergo an epithelial to mesenchymal transition, delaminating from the neuroepithelium and migrating through the periphery where they differentiate into varied cell types.[1] The emergence of neural crest was important in vertebrate evolution because many of its structural derivatives are defining features of the vertebrate clade.[2]

Underlying the development of neural crest is a gene regulatory network, described as a set of interacting signals, transcription factors, and downstream effector genes that confer cell characteristics such as multipotency and migratory capabilities.[3] Understanding the molecular mechanisms of neural crest formation is important for our knowledge of human disease because of its contributions to multiple cell lineages. Abnormalities in neural crest development cause neurocristopathies, which include conditions such as frontonasal dysplasia, Waardenburg-Shah syndrome, and DiGeorge syndrome.[1]

Therefore, defining the mechanisms of neural crest development may reveal key insights into vertebrate evolution and neurocristopathies.

30.1 History

Neural crest was first described in the chick embryo by Wilhelm His in 1868 as "the cord in between" (Zwischenstrang) because of its origin between the neural plate and non-neural ectoderm.[1] He named the tissue ganglionic crest since its final destination was each lateral side of the neural tube where it differentiated into spinal ganglia.[4] During the first half of the 20th century the majority of research on neural crest was done using amphibian embryos which was reviewed by Hörstadius (1950) in a well known monograph.[5]

Cell labeling techniques advanced the field of neural crest because they allowed researchers to visualize the migration of the tissue throughout the developing embryos. In the 1960s Weston and Chibon utilized radioisotopic labeling of the nucleus with tritiated thymidine in chick and amphibian embryo respectively. However, this method suffers from drawbacks of stability, since every time the labeled cell divides the signal is diluted. Modern cell labeling techniques such as rhodamine-lysinated dextran and the vital dye diI have also been developed to transiently mark neural crest lineages.[4]

The quail-chick marking system, devised by Nicole Le Douarin in 1969, was another instrumental technique used to track neural crest cells.[6][7] Chimeras, generated through transplantation, enabled researchers to distinguish neural crest cells of one species from the surrounding tissue of another species. With this technique, generations of scientists were able to reliably mark and study the ontogeny of neural crest cells.

30.2 Induction

A molecular cascade of events is involved in establishing the migratory and multipotent characteristics of neural crest cells. This gene-regulatory network can be subdivided into the following four sub-networks described below.

30.2.1 Inductive signals

First, extracellular signaling molecules, secreted from the adjacent epidermis and underlying mesoderm such as Wnts, BMPs and Fgfs separate the non-neural ectoderm (epidermis) from the neural plate during neural induction.[1][2]

Wnt signaling has been demonstrated in neural crest induction in several species through gain-of-function and loss-of-function experiments. In coherence with this observation, the promoter region of slug (a neural crest specific gene) contains a binding site for transcription factors involved in the activation of Wnt-dependent target genes, suggestive of a direct role of Wnt signaling in neural crest specification.[8]

The current role of BMP in neural crest formation is associated with the induction of the neural plate. BMP antagonists diffusing from the ectoderm generates a gradient of BMP activity. In this manner, the neural crest lineage forms from intermediate levels of BMP signaling required for the development of the neural plate (low BMP) and epidermis (high BMP).[1]

Fgf from the paraxial mesoderm has been suggested as a source of neural crest inductive signal. Researchers have demonstrated that the expression of dominate-negative Fgf receptor in ectoderm explants blocks neural crest induction when recombined with paraxial mesoderm.[9] Our current understanding of the role of BMP, Wnt, and Fgf pathways on neural crest specifier expression remains incomplete.

30.2.2 Neural plate border specifiers

Signaling events that establish the neural plate border lead to the expression of a set of transcription factors delineated here as neural plate border specifiers. These molecules include Zic factors, Pax3/7, Dlx5, Msx1/2 which may mediate the influence of Wnts, BMPs, and Fgfs. These genes are expressed broadly at the neural plate border region and precede the expression of bona fide neural crest markers.[2]

Experimental evidence places these transcription factors upstream of neural crest specifiers. For example, in *Xenopus* Msx1 is necessary and sufficient for the expression of Slug, Snail, and FoxD3.[10] Furthermore, Pax3 is essential for FoxD3 expression in mouse embryos.[11]

30.2.3 Neural crest specifiers

Following the expression of neural plate border specifiers is a collection of genes including Slug/Snail, FoxD3, Sox10, Sox9, AP-2 and c-Myc. This suite of genes, designated here as neural crest specifiers, are activated in emergent neural crest cells. At least in Xenopus, every neural crest specifier is necessary and/or sufficient for the expression of all other specifiers, demonstrating the existence of extensive cross-regulation.[2]

Outside of the tightly regulated network of neural crest specifiers are two other transcription factors Twist and

Id. Twist, a bHLH transcription factor, is required for mesenchyme differentiation of the pharyngeal arch structures.[12] Id is a direct target of c-Myc and is known to be important for the maintenance of neural crest stem cells.[13]

30.2.4 Neural crest effector genes

Finally, neural crest specifiers turn on the expression of effector genes, which confer certain properties such as migration and multipotency. Two neural crest effectors, *Rho GTPases* and *cadherins*, function in delamination by regulating cell morphology and adhesive properties. Sox9 and Sox10 regulate neural crest differentiation by activating many cell-type-specific effectors including Mitf, P0, Cx32, Trp and cKit.[2]

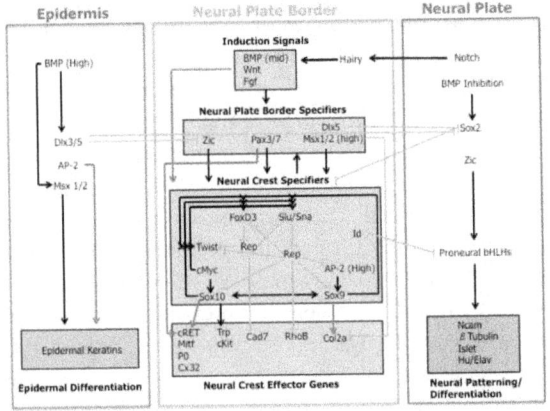

Putative neural crest gene-regulatory network functioning at the neural plate border in vertebrates. Red arrows represent proven direct regulatory interactions. Black arrows show genetic interactions based on loss-of-function and gain-of-functions studies. Gray lines denote repression. Adapted from Bronner-Fraser 2004.

30.3   Cell lineages

Neural crest cells originating from different positions along the anterior-posterior axis develop into various tissues. These regions of neural crest can be divided into four main functional domains, which include the cranial neural crest, trunk neural crest, vagal and sacral neural crest, and cardiac neural crest.

30.3.1   Cranial neural crest

Main article: cranial neural crest

Cranial neural crest migrates dorsolaterally to form the craniofacial mesenchyme that differentiates into various cranial ganglia and craniofacial cartilages and bones.[14] These cells enter the pharyngeal pouches and arches where they contribute to the thymus, bones of the middle ear and jaw and the odontoblasts of the tooth primordia.[15]

30.3.2 Trunk neural crest

Main article: trunk neural crest

Trunk neural crest gives rise to two populations of cells. One group of cells fated to become melanocytes migrates dorsolaterally into the ectoderm towards the ventral midline. A second group of cells migrates ventrolaterally through the anterior portion of each sclerotome. The cells that stay in the sclerotome form the dorsal root ganglia, whereas those that continue more ventrally form the sympathetic ganglia, adrenal medulla, and the nerves surrounding the aorta.[15]

30.3.3 Vagal and sacral neural crest

The vagal and sacral neural crest cells develop into the ganglia of the enteric nervous system and the parasympathetic ganglia.[15]

30.3.4 Cardiac neural crest

Main article: cardiac neural crest

Cardiac neural crest develops into melanocytes, cartilage, connective tissue and neurons of some pharyngeal arches. Also, this domain gives rise to regions of the heart such as the musculo-connective tissue of the large arteries, and part of the septum, which divides the pulmonary circulation from the aorta.[15] The semilunar valves of the heart are associated with neural crest cells according to new research.[16]

30.4 Evolution

Several structures that distinguish the vertebrates from other chordates are formed from the derivatives of neural crest cells. In their "New head" theory, Gans and Northcut argue that the presence of neural crest was the basis for vertebrate specific features, such as sensory ganglia and cranial skeleton. Furthermore, the appearance of these features was pivotal in vertebrate evolution because it enabled a predatory lifestyle.[17]

However, considering the neural crest a vertebrate innovation does not mean that it was created *de novo*. Instead, new structures often arise through modification of existing developmental regulatory programs. For example, regulatory programs may be changed by the co-option of new upstream regulators or by the employment of new downstream gene targets, thus placing existing networks in a novel context.[18][19] This idea is supported by in situ hybridization data that shows the conservation of the neural plate border specifiers in protochordates, which suggest that part of the neural crest precursor network was present in a common ancestor to the chordates.[3] In some non-vertebrate chordates such as tunicates a lineage of cells (melanocytes) has been identified, which are similar to neural crest cells in vertebrates. This implies that a rudimentary neural crest existed in a common ancestor of vertebrates and tunicates.[20]

30.5 Neural crest derivatives

Mesectoderm:[21] odontoblasts, dental papillae, the chondrocranium (nasal capsule, Meckel's cartilage, scleral ossicles, quadrate, articular, hyoid and columella), tracheal and laryngeal cartilage, the dermatocranium (membranous bones), dorsal fins and the turtle plastron (lower vertebrates), pericytes and smooth muscle of branchial arteries and veins, tendons of ocular and masticatory muscles, connective tissue of head and neck glands (pituitary, salivary, lachrymal, thymus, thyroid) dermis and adipose tissue of calvaria, ventral neck and face

Endocrine cells: chromaffin cells of the adrenal medulla, parafollicular cells of the thyroid, glomus cells type I/II.

Peripheral nervous system: Sensory neurons and glia of the dorsal root ganglia, cephalic ganglia (VII and in part, V, IX, and X), Rohon-Beard cells, some Merkel cells in the whisker,[22][23] Satellite glial cells of all autonomic and sensory ganglia, Schwann cells of all peripheral nerves.

Melanocytes and iris pigment cells

30.6 See also

- Neural plate

- Neural fold

- Neurulation

- DGCR2 - may control neural crest cell migration

30.7   References

[1] Huang, X., and Saint-Jeannet, J.P. (2004). "Induction of the neural crest and the opportunities of life on the edge". Dev. Biol. 275, 1-11. doi:10.1016/j.ydbio.2004.07.033

[2] Meulemans, D., and Bronner-Fraser, M. (2004). "Gene-regulatory interactions in neural crest evolution and development". Dev Cell. 7, 291-9. doi:10.1016/j.devcel.2004.08.007

[3] Sauka-Spengler, T., Meulemans, D., Jones, M., and Bronner-Fraser, M. (2007). "Ancient evolutionary origin of the neural crest gene regulatory network". Dev Cell. 13, 405-20. doi:10.1016/j.devcel.2007.08.005 PMID 17765683

[4] Le Douarin, N.M. (2004). "The avian embryo as a model to study the development of the neural crest: a long and still ongoing story". Mech Dev. 121, 1089-102. doi:10.1016/j.mod.2004.06.003

[5] Hörstadius, S. (1950). *The Neural Crest: Its Properties and Derivatives in the Light of Experimental Research.* Oxford University Press, London, 111 p.

[6] Le Douarin, N.M. (1969). "Particularités du noyau interphasique chez la Caille japonaise (Coturnix coturnix japonica). Utilisation de ces particularités comme «marquage biologique» dans les recherches sur les interactions tissulaires et les migrations cellulaires au cours de l'ontogenèse". Bull biol Fr Belg 103 : 435-52.

[7] Le Douarin, N.M. (1973). "A biological cell labeling technique and its use in experimental embryology". Dev Biol. 30 217-22. doi:10.1016/0012-1606(73)90061-4

[8] Vallin, J. et al. (2001). "Cloning and characterization of the three Xenopus slug promoters reveal direct regulation by Lef/beta-catenin signaling". J Biol Chem. 276, 30350-8. doi:10.1074/jbc.M103167200

[9] Mayor, R., Guerrero, N., Martinez, C. (1997). "Role of FGF and noggin in neural crest induction". Dev Biol. 189 1-12. doi:10.1006/dbio.1997.8634

[10] Tribulo, C. et al. (2003). "Regulation of Msx genes by Bmp gradient is essential for neural crest specification". *Development.* 130, 6441-52. doi:10.1242/dev.00878

[11] Dottori, M., Gross, M.K., Labosky, P., and Goulding, M. (2001). "The winged-helix transcription factor Foxd3 suppresses interneuron differentiation and promotes neural crest cell fate". *Development* 128, 4127–4138.

[12] Vincentz, J.W. et al. (2008). "An absence of Twist1 results in aberrant cardiac neural crest morphogenesis". Dev Biol. 320, 131-9. doi:10.1016/j.ydbio.2008.04.037

[13] Light, W. et al. (2005). "Xenopus Id3 is required downstream of Myc for the formation of multipotent neural crest progenitor cells". *Development.* 132, 1831-41. doi:10.1242/dev.01734

[14] Taneyhill, L.A. (2008). "To adhere or not to adhere: the role of Cadherins in neural crest development". Cell Adh Migr. 2, 223-30.

[15] http://www.ncbi.nlm.nih.gov/bookshelf/br.fcgi?book=dbio&part=A3109#A3133

[16] http://www.springerlink.com/content/h47w315112064434/

[17] Gans, C. and Northcutt, R. G. (1983). "Neural crest and the origin of vertebrates: A new head". *Science* 220, 268–274. doi:10.1126/science.220.4594.268

[18] Sauka-Spengler, T. and Bronner-Fraser, M. (2006). "Development and evolution of the migratory neural crest: a gene regulatory perspective". Curr Opin Genet Dev. 13, 360-6. doi:10.1016/j.gde.2006.06.006

[19] Donoghue, P.C., Graham, A., Kelsh, R.N. (2008). "The origin and evolution of the neural crest". *Bioessays.* 30, 530-41. doi:10.1002/bies.20767

[20] Abitua, P. B.; Wagner, E.; Navarrete, I. A.; Levine, M. (2012). "Identification of a rudimentary neural crest in a non-vertebrate chordate". *Nature.* doi:10.1038/nature11589.

[21] Kalcheim, C. and Le Douarin, N. M. (1998). The Neural Crest (2nd ed.). Cambridge, U. K.: Cambridge University Press.

[22] Van Keymeulen A, Mascre G, Youseff KK, Harel I, Michaux C, De Geest N, Szpalski C, Achouri Y, Bloch W, Hassan BA, Blanpain C. Epidermal progenitors give rise to Merkel cells during embryonic development and adult homeostasis. J Cell Biol. 2009 Oct 5;187(1):91-100. PubMed PMID 19786578.

[23] Szeder V, Grim M, Halata Z, Sieber-Blum M. Neural crest origin of mammalian Merkel cells. Dev Biol. 2003 Jan 15;253(2):258-63. PubMed PMID 12645929.

30.8   External links

- Embryology at UNSW *Notes/ncrest*

- ancil-445 at NeuroNames

- Diagram at University of Michigan

- Hox domains in chicks

Chapter 31

Endoderm

Endoderm is one of the three primary germ layers in the very early human embryo. The other two layers are the ectoderm (outside layer) and mesoderm (middle layer), with the endoderm being the innermost layer.[1] Cells migrating inward along the archenteron form the inner layer of the gastrula, which develops into the endoderm.

The endoderm consists at first of flattened cells, which subsequently become columnar. It forms the epithelial lining of multiple systems.

31.1 Production

The following chart shows the tissues produced by the endoderm. The embryonic endoderm develops into the interior linings of two tubes in the body, the digestive and respiratory tube.[2]

Liver and pancreas cells are believed to derive from a common precursor.[4]

In humans,the endoderm can differentiate into distinguishable organs after 5 weeks of embryonic development.

31.2 Additional images

- Section through the embryo.
- Section through ovum imbedded in the uterine decidua
- Signaling pathway to inducing endoderm

31.3 See also

- Ectoderm
- Germ layer
- Histogenesis
- Mesoderm

- Organogenesis
- Endodermal sinus tumor
- Gastrulation

31.4 References

This article incorporates text in the public domain from the 20th edition of Gray's Anatomy (1918)

[1] Langman's Medical Embryology, 11th edition. 2010.

[2] Gilbert, SF. "Endoderm". Sinauer Associates. Retrieved 14 March 2013.

[3] The **General** category denotes that all or most of the animals containing this layer produce the adjacent product.

[4] Zaret KS (October 2001). "Hepatocyte differentiation: from the endoderm and beyond". *Curr. Opin. Genet. Dev.* **11** (5): 568–74. doi:10.1016/S0959-437X(00)00234-3. PMID 11532400.

Chapter 32

Splanchnopleuric mesenchyme

In the anatomy of an embryo, the **splanchnopleuric mes-enchyme** is a structure created during embryogenesis when the lateral mesodermal germ layer splits into two layers. The inner (or splanchnic) layer adheres to the endoderm, and with it forms the splanchnopleure (mesoderm external to the coelom plus the endoderm).

32.1 See also

Post development the somato and splanchnopleuric junction lies at the duodeno-jejunal flexure.

- somatopleure

- mesenchymei

32.2 References

This article incorporates text in the public domain from the 20th edition of Gray's Anatomy (1918)

32.3 External links

- *digest-022*—Embryo Images at University of North Carolina

- *digest-023*—Embryo Images at University of North Carolina

- Overview at Kennesaw State University

Chapter 33

Mesoderm

In all bilaterian animals, the **mesoderm** is one of the three primary *germ layers* in the very early *embryo*. The other two layers are the *ectoderm* (outside layer) and *endoderm* (inside layer), with the mesoderm as the *middle* layer between them.[1][2]

The mesoderm forms mesenchyme, mesothelium, non-epithelial blood cells and coelomocytes. Mesothelium lines coeloms. Mesoderm forms the muscles in a process known as myogenesis, septa (cross-wise partitions) and mesenteries (length-wise partitions); and forms part of the gonads (the rest being the gametes).[1] Myogenesis is specifically a function of Mesenchyme.

The mesoderm differentiates from the rest of the embryo through intercellular signaling, after which the mesoderm is polarized by an organizing center.[3] The position of the organizing center is in turn determined by the regions in which beta-catenin is protected from degradation by GSK-3. Beta-catenin acts as a co-factor that alters the activity of the transcription factor tcf-3 from repressing to activating, which initiates the synthesis of gene products critical for mesoderm differentiation and gastrulation. Furthermore, mesoderm has the capability to induce the growth of other structures, such as the neural plate, the precursor to the nervous system.

33.1 Definition

The mesoderm is one of the three germinal layers that appears in the third week of embryonic development. It is formed through a process called gastrulation. There are three important components, the paraxial mesoderm, the intermediate mesoderm and the lateral plate mesoderm. The paraxial mesoderm forms the somitomeres, which give rise to mesenchyme of the head and organize into somites in occipital and caudal segments. Somites give rise to the myotome (muscle tissue), sclerotome (cartilage and bone), and dermatome (subcutaneous tissue of the skin).[1][2] Signals for somite differentiation are derived from surroundings structures, including the notochord, neural tube and epidermis. The intermediate mesoderm connects the paraxial mesoderm with the lateral plate, eventually it differentiates into urogenital structures consisting of the kidneys, gonads, their associated ducts, and the adrenal glands. The lateral plate mesoderm give rise to the heart, blood vessels and blood cells of the circulatory system as well as to the mesodermal component of the limbs.[4] Some of the mesoderm derivatives include the muscle (smooth, cardiac and skeletal), the muscles of the tongue (occipital somites), the pharyngeal arches muscle (muscles of mastication, muscles of facial expressions), connective tissue, dermis and subcutaneous layer of the skin, bone and cartilage, dura mater, endothelium of blood vessels, red blood cells, white blood cells, microglia and Kupffer cells, the kidneys and the adrenal cortex.[5]

33.2 Development of the mesodermal germ layer

During the third week a process called gastrulation creates a mesodermal layer between the endoderm and the ectoderm. This process begins with formation of a primitive streak on the surface of the epiblast.[6] The cells of the layers move between the epiblast and hypoblast and begin to spread laterally and cranially. The cells of the epiblast move toward the primitive streak and slip beneath it in a process called invagination. Some of the migrating cells displace the hypoblast and create the endoderm, and others migrate between the endoderm and the epiblast to create the mesoderm. The remaining cells form the ectoderm. After that, the epiblast and the hypoblast establish contact with the extraembryonic mesoderm until they cover the yolk sac and amnion. They move onto either side of the prechordal plate. The prechordal cells migrate to the midline to form the notochordal plate. The chordamesoderm is the central region of trunk mesoderm.[4] This forms the notochord which induces the formation of the neural tube and establishes the anterior-posterior body axis. The notochord extends beneath the neural tube from the head to the tail. The mesoderm moves to the midline until it covers the notochord, when the mesoderm cells proliferate they form the paraxial mesoderm. In each side, the mesoderm remains thin and is known as the lateral plate. The intermediate mesoderm lies between the paraxial mesoderm and the lateral plate. Between days 13 and 15, the proliferation of extraembryonic mesoderm, primitive streak and embryonic mesoderm take place. The notochord process occurs between days 15 and 17. Eventually, the development of the notochord canal and the axial canal takes place between days 17 and 19 when the first three somites are formed.[7]

33.3   Paraxial mesoderm

During the third week, the paraxial mesoderm is organized into segments. If they appear in the cephalic region and grow with cephalocaudal direction, they are called somitomeres. If they appear in the cephalic region but establish contact with the neural plate, they are known as neuromeres, which later will form the mesenchyme in the head. The somitomeres organize into somites which grow in pairs. In the fourth week the somites lose their organization and cover the notochord and spinal cord to form the backbone. In the fifth week, there are 4 occipital somites, 8 cervical, 12 thoracic, 5 lumbar, 5 sacral and 8 to 10 coccygeal that will form the axial skeleton. Somatic derivatives are determined by local signaling between adjacent embryonic tissues, in particular the neural tube, notochord, surface ectoderm and the somatic compartments themselves.[8] The correct specification of the deriving tissues, skeletal, cartilage, endothelia and connective tissue is achieved by a sequence of morphogenic changes of the paraxial mesoderm, leading to the three transitory somatic compartments: dermomyotome, myotome and sclerotome. These structures are specified from dorsal to ventral and from medial to lateral.[8] each somite will form its own sclerotome that will differentiate into the tendon cartilage and bone component. Its myotome will form the muscle component and the dermatome that will form the dermis of the back. The myotome and dermatome have a nerve component.[1][2]

33.4   Molecular Regulation of Somite Differentiation

Surrounding structures such as the notochord, neural tube, epidermis and lateral plate mesoderm send signals for somite differentiation[1][2] Notochord protein accumulates in presomitic mesoderm destined to form the next somite and then decreases as that somite is established. The notochord and the neural tube activate the protein SHH which

helps the somite to form its sclerotome. The cells of the sclerotome express the protein PAX1 that induces the cartilage and bone formation. The neural tube activates the protein WNT1 that expresses PAX 2 so the somite creates the myotome and dermatome. Finally, the neural tube also secretes neurotrophin 3 (NT-3), so that the somite creates the dermis. Boundaries for each somite are regulated by retinoic acid (RA) and a combination of FGF8and WNT3a.[1][2] So the retinoic acid is and endogenous signal that maintains the bilateral synchrony of mesoderm segmentation and controls bilateral symmetry in vertebrates. The bilaterally symmetric body plan of vertebrate embryos is obvious in somites and their derivates such as the vertebral column. Therefore, asymmetric somite formation correlates with a left-right desynchronization of the segmentation oscillations.[9]

Many studies with Xenopus and zebrafish have analyzed the factors of this development and how they interact in signaling and transcription. However, there are still some doubts in how the prospective mesodermal cells integrate the various signals they receive and how they regulate their morphogenic behaviours and cell-fate decisions.[8] Human embryonic stem cells for example have the potential to produce all of the cells in the body and they are able to self-renew indefinitely so they can be used for a large-scale production of therapeutic cell lines. They are also able to remodel and contract collagen and were induced to express muscle actin. This shows that these cells are multipotent cells.[10]

33.5   Intermediate mesoderm

The intermediate mesoderm connects the paraxial mesoderm with the lateral plate and differentiates into urogenital structures.[11] In upper thoracic and cervical regions this forms the nephrotomes, and in caudally regions this forms the nephrogenic cord. It also helps to develop the excretory units of the urinary system and the gonads.[4]

33.6   Lateral plate mesoderm

The lateral plate mesoderm splits into parietal (somatic) and visceral (splanchnic) layers. The formation of these layers starts with the appearance of intercellular cavities.[11] The somatic layer depends on a continuous layer with mesoderm that covers the amnion. The splanchnic depends on a continuous layer that covers the yolk sac. The two layers cover the intraembryonic cavity. The parietal layer together with overlying ectoderm forms the lateral body wall folds. The visceral layer forms the walls of the gut tube. Mesoderm cells of the parietal layer form the mesothelial membranes or serous membranes which line the peritoneal, pleural and pericardial cavities.[1][2]

33.7   See also

- Chordamesoderm (also known as *axial mesoderm*) which later on gives rise to notochord in all chordates
- Embryogenesis
- Gastrulation
- Histogenesis
- Intermediate mesoderm
- Lateral plate mesoderm
- Mesenchyme
- Mesothelium
- Organogenesis
- Paraxial mesoderm
- Somites
- Triploblastic

33.8   References

[1] Ruppert, E.E., Fox, R.S., and Barnes, R.D. (2004). "Introduction to Bilateria". *Invertebrate Zoology* (7th ed.). Brooks/Cole. pp. 217–218. ISBN 0-03-025982-7.

[2] Langman's Medical Embryology, 11th edition. 2010.

[3] Kimelman, D. & Bjornson, C. (2004). "Vertebrate Mesoderm Induction: From Frogs to Mice". In Stern, Claudio D. *Gastrulation: from cells to embryo*. CSHL Press. p. 363. ISBN 978-0-87969-707-5.

[4] Scott, Gilbert (2010). *Developmental biology* (ninth ed.). USA: Sinauer Associates.

[5] Dudek, Ronald W. (2009). *High-yield. Embryology* (4th ed.). Lippincott Williams & Wilkins.

[6] "Paraxial Mesoderm: The somites and their derivatives". NCBI. Retrieved April 15, 2013.

[7] Drew, Ulrich (1993). *Color atlas of embryology*. German: Thieme.

[8] Yusuf, Faisal (2006). "The eventful somite: Patterning, fate determination and cell division in the somite". *Anatomy and embryology*: 21–30.

[9] Vermot, J.; Gallego Llamas, J.; Fraulob, V.; Niederrei-ther, K.; Chambon, P.; Dollé, P. (April 2005). "Retinoic acid controls the bilateral symmetry of somite formation in the mouse embryo". *Science* **308** (5721): 563–566. doi:10.1126/science.1108363. PMID 15731404.

[10] Boyd, N.L.; Robbins KR, K.R.; Dhara SK, S.K.; West FD,, F.D.; Stice SL., S.L. (August 2009). "Human embryonic stem cell-derived mesoderm-like epithelium transitions to mesenchymal progenitor cells". *Tissue Engineering. Part A.* **15** (8): 1897–1907. doi:10.1089/ten.tea.2008.0351. PMID 19196144.

[11] Kumar, Rani (2008). *Textbook of human embryology*. I.K. International.

33.9   Further reading

- Gurdon, J.B. (1995). "The formation of mesoderm and muscle in *Xenopus*". In Zagris, Nikolas; et al. *Organization of the early vertebrate embryo*. Springer. ISBN 978-0-306-45132-4.

- Kenderew, John Cowdery & Lawrence, Eleanor, eds. (1994). "Mesoderm Induction". *The encyclopedia of molecular biology*. John Wiley & Sons. p. 541. ISBN 978-0-632-02182-6.

- Liu, Shu Q. (2007). "Early Embryonic Organ Development". *Bioregenerative engineering: principles and applications*. John Wiley & Sons. ISBN 978-0-471-70907-7.

- McGeady, Thomas A.; et al. (2006). "Establishment of the Basic Body Plan". *Veterinary embryology*. Wiley-Blackwell. ISBN 978-1-4051-1147-8.

- Pappaioannou, Virginia, E. (2004). "Early Embryonic Mesoderm Development". In Lanza, Robert Paul. *Handbook of stem cells, Volume 1*. Gulf Professional Publishing. ISBN 978-0-12-436642-8.

- Sherman, Lawrence S. et al., eds. (2001). *Human embryology* (3rd ed.). Elsevier Health Sciences. ISBN 978-0-443-06583-5.

33.10   External links

- Embryology at UNSW *Notes/skmus6*
- Embryology at Temple *EMBIII97/sld039*

Chapter 34

Axial mesoderm

Axial mesoderm, or **chordamesoderm**, is a type of mesoderm that lies along the central axis under the neural tube.

- will give rise to notochord

- starts as the notochordal process, whose formation finishes at day 20.

- important not only in forming the notochord itself but also in inducing development of the overlying ectoderm into the neural tube

- will eventually induce the formation of vertebral bodies.

- ventral floor of the notochordal process fuses with endoderm.

- The notochord will form the nucleus pulposus of interverterbral discs. There is some discussion as to whether these cells contributed from the notochord are replaced by others from the adjacent mesoderm.

It gives rise to the notochordal process, which later becomes the notochord.

34.1 References

This article incorporates text in the public domain from the 20th edition of Gray's Anatomy (1918)

34.2 External links

- Diagrams at cornell.edu

Chapter 35

Paraxial mesoderm

Paraxial mesoderm, also known as **presomitic** or **somitic mesoderm** is the area of mesoderm in the neurulating embryo that flanks and forms simultaneously with the neural tube. The cells of this region give rise to somites, blocks of tissue running along both sides of the neural tube, which form muscle and the tissues of the back, including connective tissue and the dermis.

35.1   Formation and somitogenesis

The paraxial and other regions of the mesoderm are thought to be specified by bone morphogenetic proteins, or BMPs, along an axis spanning from the center to the sides of the body. Members of the FGF family also play an important role, as does the WNT pathway. In particular, Noggin, a downstream target of the Wnt pathway, antagonizes BMP signaling, forming boundaries where antagonists meet and limiting this signaling to a particular region of the mesoderm. Together, these pathways provide the initial specification of the paraxial mesoderm and maintain this identity.[1] This specification process has now been fully recapitulated *in vitro* with the formation of paraxial mesoderm progenitors from pluripotent stem cells, using a directed differentiation approach.[2]

The tissue undergoes convergent extension as the primitive streak regresses, or as the embryo gastrulates. The notochord extends from the base of the head to the tail; with it extend thick bands of paraxial mesoderm.[3]

As the primitive streak continues to regress, somites form from the paraxial mesoderm by "budding off" rostrally.

In certain model systems, it has been shown that the daughter cells of stem cell-like progenitor cells which come from the primitive streak or site of gastrulation migrate out and localize in the posterior paraxial mesoderm. As the primitive streak regresses and somites bud off anteriorly, new cells derived from these stem-cell like precursors constantly enter the posterior end of the paraxial mesoderm.[4][5]

35.2   Derived tissues

Many kinds of tissue derive from the segmented paraxial mesoderm by means of the somite. Among these are:

- the sclerotome, which forms cartilage,

- the syndotome, which forms tendons,

- the myotome, which forms skeletal muscle,

- the dermatome, which forms the dermis as well as skeletal muscle,

- and endothelial cells.

35.2.1   Head Mesoderm

A particular kind of tissue deriving from the paraxial mesoderm is the head mesoderm. This tissue derives from the unsegmented paraxial mesoderm and prechordal mesoderm. Tissues derived from the head mesoderm include connective tissue and the muscles of the face.

The head mesoderm forms through a separate signaling circuit than the segmented paraxial mesoderm, though also involving BMP and fibroblast growth factor (FGF) signaling. Here, retinoic acid interacts with these pathways.[6]

35.3   See also

- Somitomere

- Chordamesoderm

- Intermediate mesoderm

- Lateral plate mesoderm

- Mesenchyme

- Triploblasty

35.4   References

This article incorporates text in the public domain from the 20th edition of Gray's Anatomy (1918)

[1] Pourquié, O. (2001). "Vertebratesomitogenesis". *Annual Review of Cell and Developmental Biology* **17**: 311–350. doi:10.1146/annurev.cellbio.17.1.311. PMID 11687492.

[2] Chal J, Oginuma M, Al Tanoury Z, Gobert B, Sumara O, Hick A, Bousson F, Zidouni Y, Mursch C, Moncuquet P, Tassy O, Vincent S, Miyanari A, Bera A, Garnier JM, Guevara G, Hestin M, Kennedy L, Hayashi S, Drayton B, Cherrier T, Gayraud-Morel B, Gussoni E, Relaix F, Tajbakhsh S, Pourquié O (August 2015). "Differentiation of pluripotent stem cells to muscle fiber to model Duchenne muscular dystrophy". *Nature Biotechnology.* doi:10.1038/nbt.3297. PMID 26237517.

[3] Gilbert, S.F. (2010). *Developmental Biology* (9th ed.). Sinauer Associates, Inc. pp. 413–415. ISBN 978-0-87893-384-6.

[4] Cambray, N.; Wilson, V. (2007). "Two distinct sources for a population of maturing axial progenitors". *Development* **134** (15): 2829–2840. doi:10.1242/dev.02877. PMID 17611225.

[5] Maroto, M.; Bone, R. A.; Dale, J. K. (2012). "Somitogenesis". *Development* **139** (14): 2453–2456. doi:10.1242/dev.069310. PMID 22736241.

[6] Bothe, I.; Tenin, G.; Oseni, A.; Dietrich, S. (2011). "Dynamic control of head mesoderm patterning". *Development* **138** (13): 2807–2821. doi:10.1242/dev.062737. PMID 21652653.

35.5   External links

- Overview at nih.gov
- Diagram and overview at mcgill.ca
- Diagram at lww.com

Chapter 36

Somite

A **somite** is a division of the body of an animal or embryo.

Somites are bilaterally paired blocks of paraxial mesoderm that form along the head-to-tail axis of the developing embryo in segmented animals. In vertebrates, somites subdivide into the sclerotomes, myotomes and dermatomes that give rise to the vertebrae of the vertebral column, rib cage, and part of the occipital bone; skeletal muscle, cartilage, tendons, and skin (of the back).[2]

The word *somite* is also used in place of the word *metamere*. In this definition, the somite is a homologously paired structure in an animal body plan, such as is visible in annelids and arthropods.[3]

36.1 Development

The mesoderm forms at the same time as the other two germ layers, the ectoderm and endoderm. The mesoderm at either side of the neural tube is called paraxial mesoderm. It is distinct from the mesoderm underneath the neural tube which is called the chordamesoderm that becomes the notochord. The paraxial mesoderm is initially called the "segmental plate" in the chick embryo or the "unsegmented mesoderm" in other vertebrates. As the primitive streak regresses and neural folds gather (to eventually become the neural tube), the paraxial mesoderm separates into blocks called somites.[4]

36.1.1 Formation

The pre-somitic mesoderm commits to the somitic fate before mesoderm becomes capable of forming somites. The cells within each somite are specified based on their location within the somite. Additionally, they retain the ability to become any kind of somite-derived structure until relatively late in the process of somitogenesis.[4]

The development of the somites depends on a clock mechanism as described by the clock and wavefront model. In one description of the model, oscillating Notch and Wnt

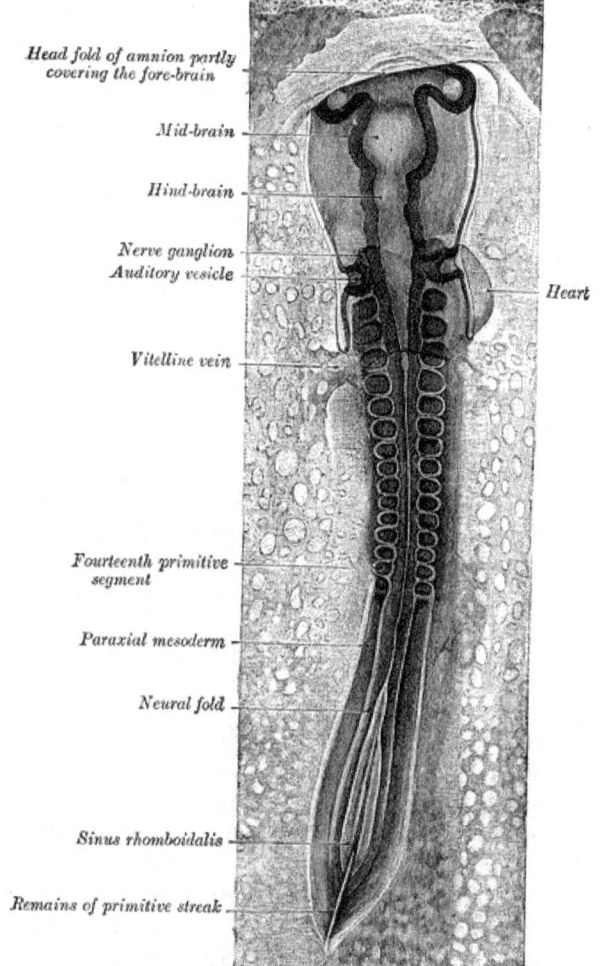

Head fold of amnion partly covering the fore-brain

Mid-brain

Hind-brain

Nerve ganglion
Auditory vesicle

Heart

Vitelline vein

Fourteenth primitive segment

Paraxial mesoderm

Neural fold

Sinus rhomboidalis

Remains of primitive streak

Chick embryo of thirty-three hours' incubation, viewed from the dorsal aspect. X 30

signals provide the clock. The wave is a gradient of the FGF protein that is rostral to caudal (nose to tail gradient). Somites form one after the other down the length of the embryo from the head to the tail, with each new somite forming on the caudal (tail) side of those already in existing

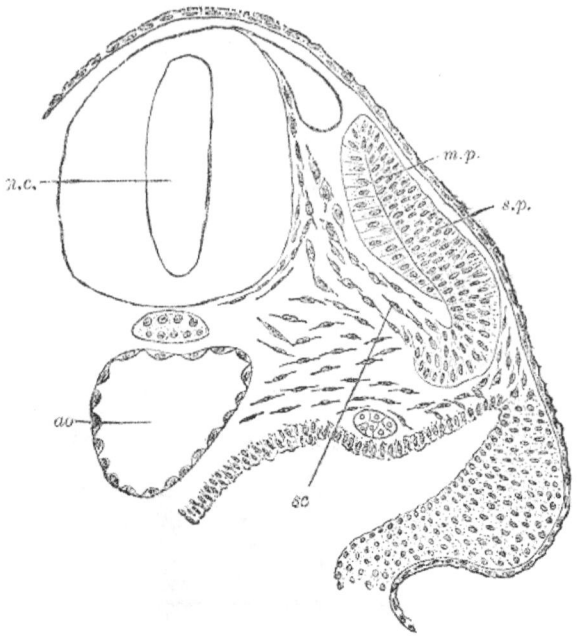

Transverse section of a human embryo of the third week to show the differentiation of the primitive segment. ao. Aorta. m.p. Muscle-plate. n.c. Neural canal. sc. Sclerotome. s.p. Dermatome

Notch signalling

The *Notch* system, as part of the clock and wavefront model, forms the boundaries of the somites. *DLL1* and *DLL3* are *Notch* ligands, mutations of which cause various defects. Notch regulates *HES1*, which sets up the caudal half of the somite. *Notch* activation turns on *LFNG* which in turn inhibits the *Notch* receptor. *Notch* activation also turns on the *HES1 gene* which inactivates *LFNG*, re-enabling the *Notch* receptor, and thus accounting for the oscillating clock model. *MESP2* induces the *EPHA4* gene, which causes repulsive interaction that separates somites by causing segmentation. *EPHA4* is restricted to the boundaries of somites. *EPHB2* is also important for boundaries.

Mesenchymal-epithelial transition

Fibronectin and N-cadherin are key to the mesenchymal-epithelial transition process in the developing embryo. The process is probably regulated by paraxis and *MESP2*. In turn, *MESP2* is regulated by *Notch* signaling. Paraxis is regulated by processes involving the cytoskeleton.

Specification

somites.[5][5][6]

The timing of the interval is not universal. Different species have different interval timing. In the chick embryo somites are formed every 90 minutes. In the mouse the interval is variable.

For some species, the number of somites may be used to determine the stage of embryonic development more reliably than the number of hours post-fertilization because rate of development can be affected by temperature or other environmental factors. The somites appear on both sides of the neural tube simultaneously. Experimental manipulation of the developing somites will not alter the rostral/caudal orientation of the somites, as the cell fates have been determined prior to somitogenesis. Somite formation can be induced by *Noggin*-secreting cells. The number of somites is species dependent and independent of embryo size (for example, if modified via surgery or genetic engineering). Chicken embryos have 50 somites; mice have 65, while snakes have 500.[4][7]

As cells within the paraxial mesoderm begin to come together, they are termed somitomeres, indicating a lack of complete separation between segments. The outer cells undergo a mesenchymal–epithelial transition to form an epithelium around each somite. The inner cells remain as mesenchyme.

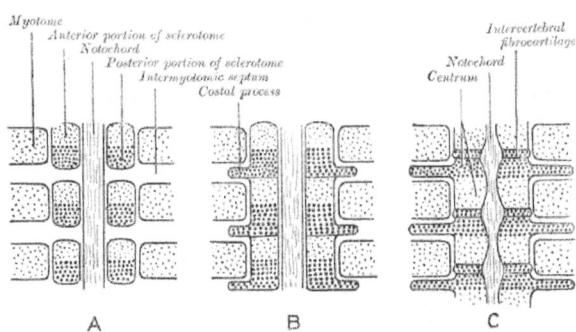

Scheme showing how each vertebral centrum is developed from portions of two adjacent segments. (Myotome labelled in upper left.)

The Hox genes specify somites as a whole based on their position along the anterior-posterior axis through specifying the pre-somitic mesoderm before somitogenesis occurs. After somites are made, their identity as a whole has already been determined, as is shown by the fact that transplantation of somites from one region to a completely different region results in the formation of structures usually observed in the original region. In contrast, the cells within each somite retain plasticity (the ability to form any kind of structure) until relatively late in somitic development.[4]

36.1.2 Derivatives

In the developing vertebrate embryo, somites split to form dermatomes, skeletal muscle (myotomes), tendons and cartilage (syndetomes)[8] and bone (sclerotomes).

Because the sclerotome differentiates before the dermatome and the myotome, the term *dermomyotome* refers to the combined dermatome and myotome before they too separate out.[9]

Dermatome

The **dermatome** is the dorsal portion of the paraxial mesoderm somite which gives rise to the skin (dermis). In the human embryo it arises in the third week of embryogenesis.[2] It is formed when a dermamyotome (the remaining part of the somite left when the sclerotome migrates), splits to form the dermatome and the myotome.[2] The dermatomes contribute to the skin, fat and connective tissue of the neck and of the trunk, though most of the skin is derived from lateral plate mesoderm.[2]

Myotome

The **myotome** is that part of a somite that forms the muscles of the animal.[2] Each myotome divides into an epaxial part (*epimere*), at the back, and a hypaxial part (*hypomere*) at the front.[2] The myoblasts from the hypaxial division form the muscles of the thoracic and anterior abdominal walls. The epaxial muscle mass loses its segmental character to form the extensor muscles of the neck and trunk of mammals.

In fishes, salamanders, caecilians, and reptiles, the body musculature remains segmented as in the embryo, though it often becomes folded and overlapping, with epaxial and hypaxial masses divided into several distinct muscle groups.

Sclerotome

The **sclerotome** forms the vertebrae and the rib cartilage and part of the occipital bone; the myotome forms the musculature of the back, the ribs and the limbs; the syndetome forms the tendons and the dermatome forms the skin on the back. In addition, the somites specify the migration paths of neural crest cells and the axons of spinal nerves. From their initial location within the somite, the sclerotome cells migrate medially towards the notochord. These cells meet the sclerotome cells from the other side to form the vertebral body. The lower half of one sclerotome fuses with the upper half of the adjacent one to form each vertebral body.[10] From this vertebral body, sclerotome cells move dorsally and surround the developing spinal cord, forming the vertebral arch. Other cells move distally to the costal processes of thoracic vertebrae to form the ribs.[10]

36.2 In arthropods

In crustacean development, a somite is a segment of the hypothetical primitive crustacean body plan. In current crustaceans, several of those somites may be fused.

36.3 In older texts

In some older texts, somites are referred to as "primitive segments."

36.4 See also

- Developmental biology

36.5 References

[1] "The Third Week Of Life:". Retrieved 2007-10-13.

[2] Larsen, William J. (2001). *Human embryology* (3. ed.). Philadelphia, Pa.: Churchill Livingstone. pp. 53–86. ISBN 0-443-06583-7.

[3] "Metamere". *Dictionary and Thesaurus-Merriam-Webster Online*. Merriam-Webster. 2012. Retrieved 11 December 2012.

[4] Gilbert, S.F. (2010). *Developmental Biology* (9th ed.). Sinauer Associates, Inc. pp. 413–415. ISBN 978-0-87893-384-6.

[5] Baker, R. E.; Schnell, S.; Maini, P. K. (2006). "A clock and wavefront mechanism for somite formation". *Developmental Biology* **293** (1): 116–126. doi:10.1016/j.ydbio.2006.01.018. PMID 16546158.

[6] Goldbeter, A.; Pourquié, O. (2008). "Modeling the segmentation clock as a network of coupled oscillations in the Notch, Wnt and FGF signaling pathways". *Journal of Theoretical Biology* **252** (3): 574–585. doi:10.1016/j.jtbi.2008.01.006. PMID 18308339.

[7] Gomez, C; et al. (2008). "Control of segment number in vertebrate embryos". *Nature* **454** (7202): 335–339. doi:10.1038/nature07020. PMID 18563087.

[8] Brent AE, Schweitzer R, Tabin CJ (April 2003). "A somitic compartment of tendon progenitors". *Cell* **113** (2): 235–48. doi:10.1016/S0092-8674(03)00268-X. PMID 12705871. Retrieved 2014-04-20.

[9] "med.unc.edu". Retrieved 2007-10-19.

[10] Walker, Warren F., Jr. (1987) *Functional Anatomy of the Vertebrate* San Francisco: Saunders College Publishing.

36.6   External links

- Embryology at UNSW *Notes/week3_6*

Chapter 37

Somitomere

In the developing vertebrate embryo, the **somitomeres** (or **somatomeres**)[1] are cells that are derived from the loose masses of paraxial mesoderm that are found alongside the developing neural tube. In human embryogenesis they appear towards the end of the third gestational week. The approximately 50 pairs of somitomeres in the human embryo, begin developing in the cranial (head) region, continuing in a caudal (tail) direction until the end of week four.

37.1 Development

The first seven somitomeres give rise to the striated muscles of the face, jaws, and throat.[2]

The remaining somitomeres, likely driven by periodic expression of the *hairy* gene, begin expressing adhesion proteins such as N-cadherin and fibronectin, compact, and bud off forming somites. The somites give rise to the vertebral column (sclerotome), associated muscles (myotome), and overlying dermis (dermatome). There are a total of 37 somite pairs at the end of the fifth week of development, after the first occipital somite and 5-7 coccygeal somites disappear from the original 42-44 somites

37.2 References

[1] Antonio Nanci (2008). *Ten Cate's oral histology: development, structure, and function*. Elsevier Health Sciences. pp. 25–. ISBN 978-0-323-04557-5. Retrieved 16 April 2010.

[2] Larsen W.J. Human Embryology. Churchill Livingstone.Third edition 2001.Page 62. ISBN 0-443-06583-7

37.3 External links

- Swiss embryology (from UL, UB, and UF) *hdisqueembry/triderm07* - look for Fig. 16

Chapter 38

Intermediate mesoderm

Intermediate mesenchyme or **intermediate mesoderm** is a type of mesoderm (an embryological tissue) that is located between the paraxial mesoderm and the lateral plate.

It develops into the part of the urogenital system (kidneys and gonads), as well as the reproductive system.

- forms of urogenital system

- series of short evaginations from each segment grows dorsally caudally

- vestiges of the future kidney, the pronephros briefly appears.

- pronephric duct arises in the intermediate mesoderm just ventral to the anterior somites

- grows caudally until it becomes the cloaca.

- it is distinct from the lateral mesoderm, as it is not influenced by the secretion of BMP-4 by the ectoderm, possibly due to the lack of receptors.

- Diagram at lww.com

- Overview at nih.gov

38.1 See also

- mesenchyme

- Intermediate Mesoderm of Swartz

38.2 References

This article incorporates text in the public domain from the 20th edition of Gray's Anatomy (1918)

38.3 External links

- *genital-002*—Embryo Images at University of North Carolina

Chapter 39

Lateral plate mesoderm

Lateral plate mesoderm is a type of mesoderm that is found at the periphery of the embryo.

39.1 Division into layers

It will split into two layers, the somatic layer/mesoderm and the splanchnic layer/mesoderm

- The *somatopleuric layer* forms the future body wall.

- The *splanchnopleuric layer* forms the circulatory system.

Spaces within the lateral plate are enclosed and forms the intraembryonic coelom.

It is formed by the secretion of BMP-4 by the ectoderm.[1]

39.2 Serosal mesoderms

Lateral plate mesoderm gives rise to the serosal mesoderms.[2]

- forms a ventral layer associated with endoderm, the splanchnopleuric mesoderm. This forms the viscera and heart
- forms a dorsal layer associated with ectoderm, the somatopleuric mesoderm. This forms the body wall lining and dermis.
- Abdominal portion becomes contained in dorsal mesentery, part of the serosal mesoderm.
- When the two layers form, a cardiogenic plate is visible. Later,

this will form the myocardial primordium, which will contribute to the tubular heart.

39.3 Cavities

In the 4th week the coelom divides into pericardial, pleural and peritoneal cavities.[2]

- First partition: is the septum transversum.

  - This will be translocated later into the diaphragm and ventral mesentery.

  - Divides the coelom into primitive pericardial and peritoneal cavities

- Pleuropericardial folds appear on the lateral wall of primitive pericardial cavity, which will eventually cause a partition to form the pericardial and pleural cavities.

- Communication between these partitions formed by the pericardioperitoneal canals. However, pleuroperitoneal membranes will grow to fuse with the septum transversum to close off these canals.

- At day 22, lung buds form, remaining ensheathed in a splanchnopleuric mesoderm

39.4   Limb Development

Cells from the lateral plate mesoderm and the myotome migrate to the limb field and proliferate to create the limb bud. The lateral plate cells produce the cartilaginous and skeletal portions of the limb while the myotome cells produce the muscle components. The lateral plate mesodermal cells secrete a fibroblast growth factor (FGF7 and FGF10, presumably) to induce the overlying ectoderm to form an important organizing structure called the apical ectodermal ridge (AER).The AER reciprocatively secretes FGF8 and FGF4 which maintains the FGF10 signal and induces proliferation in the mesoderm. The position of FGF10 expression is regulated by Wnt8c in the hindlimb and Wnt2b in the forelimb. The forelimb and the hindlimb are specified by their position along the anterior/posterior axis and possibly by two T-box containing transcription factors: Tbx5 and Tbx4, respectively.

39.5   See also

- Limb development for more information

- Serous membrane

39.6   References

This article incorporates text in the public domain from the 20th edition of Gray's Anatomy (1918)

[1] Tonegawa A, Funayama N, Ueno N, Takahashi Y (1997). "Mesodermal subdivision along the mediolateral axis in chicken controlled by different concentrations of BMP-4". *Development* **124** (10): 1975–84. PMID 9169844.

[2] Larsen, William J. (1998). *Essentials of human embryology.* Edinburgh: Churchill Livingstone. ISBN 0-443-07514-X.

39.7   External links

- Swiss embryology (from UL, UB, and UF) *hdisqueembry/triderm08*

Chapter 40

Intra-embryonic coelom

In the development of the human embryo the **intraembryonic coelom** (or **somatic coelom**) is a portion of the conceptus forming in the mesoderm during the third week of development. During the third week of development, the lateral mesoderm splits into a dorsal somatic mesoderm (somatopleure) and a ventral splanchnic mesoderm (splanchnopleure). The resulting cavity between the somatopleure and splanchnopleure is called the **intraembryonic coelom**. This space will give rise to the thoracic and abdominal cavities. The coelomic spaces in the lateral mesoderm and cardiogenic area are isolated. The isolated coelom begin to organize into a horseshoe shape. The spaces soon join together and form a single horseshoe-shaped cavity, the Intraembryonic Coelom which separates the mesoderm into two layers

It briefly has a connection with the extraembryonic coelom.

40.1 External links

- http://www.embryology.ch/anglais/hdisqueembry/triderm09.html

- http://embryology.med.unsw.edu.au/Notes/coelom.htm

- http://www.ana.ed.ac.uk/database/humat/notes/embryo/cavities/coelom.htm

40.2   Text and image sources, contributors, and licenses

40.2.1   Text

- **Human embryogenesis** *Source:* https://en.wikipedia.org/wiki/Human_embryogenesis?oldid=710749080 *Contributors:* AxelBoldt, Gracefool, Beland, Bender235, Arcadian, Wouterstomp, Seans Potato Business, Woohookitty, Rjwilmsi, Kolbasz, Bgwhite, Wavelength, Hairy Dude, Mfero, Dhollm, JPMcGrath, SmackBot, Krychek, Lagringa, Zephyris, Andrew c, JHunterJ, Kudakups, Was a bee, Svna91, Casliber, James086, OckRaz, WhatamIdoing, NikNaks, CFCF, Zoara, Boghog, Mikael Häggström, MishaPan, Squids and Chips, Signalhead, TXiKiBoT, Lova Falk, Northfox, Doovie, Plynch22, Ascidian, Maxcip, Niceguyedc, WikHead, Tiphaine800, Addbot, Anxietycello, Yobot, AnomieBOT, Nishanthb, Z0OMD, Machn, I dream of horses, Rayman60, Gongoozler123, Wafaashohdy, ClueBot NG, BG19bot, ManyueGPH, ChrisGualtieri, GoShow, Makecatbot, Zumwalte, MartianCat, Iztwoz, NJSfour, Sharif uddin, Monkbot, CV9933, Abirnehal, PrismTheDragon and Anonymous: 40

- **Human fertilization** *Source:* https://en.wikipedia.org/wiki/Human_fertilization?oldid=707383449 *Contributors:* Halfdan, Andycjp, R. fiend, Yik Lin Khoo, Anythingyouwant, Freakofnurture, Bender235, ESkog, Arcadian, Alansohn, Evil Monkey, Kitch, Rjwilmsi, Tdowling, AySz88, RobyWayne, DVdm, Sceptre, Rsrikanth05, Mark Kim, Brian Crawford, Allens, Lyrl, Crystallina, SmackBot, Ashill, Jfurr1981, Gilliam, Hmains, Amatulic, Can't sleep, clown will eat me, Flyguy649, AngelSL, Shinryuu, Spook`, Ginkgo100, Missionary, Fezz, Bazzargh, DumbBOT, Krm500, Plateblock, Northumbrian, Fr33ke, Seaphoto, Mack2, One Artiste, MikeLynch, Ctill, Bongwarrior, QuizzicalBee, Homunq, WhatamIdoing, LookingGlass, Canyouhearmenow, DerHexer, Tyler path, Pruthvi.Vallabh, CommonsDelinker, JeremyWJ, Mikael Häggström, Kyle the bot, Philip Trueman, Tameeria, Anna Lincoln, Donarnold, Lova Falk, Temporaluser, Darthsion101, Bfpage, Maddiekate, Flyer22 Reborn, RandomHumanoid, Plynch22, Loren.wilton, ClueBot, The Thing That Should Not Be, Arakunem, Auntof6, Excirial, KSUdvm2b, Helenginn, KeasbeyMornings, Daughter of Mímir, MystBot, Addbot, Non-dropframe, Ronhjones, Lonely goatherd, Abiyoyo, Gail, Jarble, Angrysockhop, Yobot, Fraggle81, AnakngAraw, Fatal!ty, Materialscientist, The High Fin Sperm Whale, ArthurBot, Glenk1973, Capricorn42, Papercutbiology, Shadowjams, FrescoBot, ThiagoRuiz, Pinethicket, I dream of horses, Rushbugled13, Pmokeefe, Ryt 007, Merlion444, Curiousranger, Aslmolin, John of Reading, Theonlymichael, Dcirovic, RedChuck14, Openstrings, Brandmeister, Ready, LibertyOrDeath, Judygreenberg, Wafaashohdy, ClueBot NG, Joezak, Ctap, Widr, Helpful Pixie Bot, Slefuyhiluaenrflouhesf, Lowercase sigmabot, AdventurousSquirrel, MrBill3, Dexbot, FamAD123, Sungta, AioftheStorm, Ginsuloft, BruceBlaus, Keyboardthegreater, Monkbot, Oiyarbepsy, Shadowhunterxd, Anorhanian, JumpiMaus and Anonymous: 194

- **Oocyte activation** *Source:* https://en.wikipedia.org/wiki/Oocyte_activation?oldid=692555789 *Contributors:* Bender235, Rjwilmsi, Mikael Häggström, Bfpage, Ljaffe, Addbot, Yobot, Ripchip Bot, Mesoderm, Betterthanfigs, Dexbot, Sophiepickles, Inzaie, Ebateson and Anonymous: 5

- **Zygote** *Source:* https://en.wikipedia.org/wiki/Zygote?oldid=711191923 *Contributors:* Kpjas, Mav, Tarquin, Css, PierreAbbat, LionKimbro, Icarus~enwiki, Lexor, Evercat, Ehn, Marshman, Samsara, Toreau, Robbot, Nurg, Diberri, Alan Liefting, Marc Venot, Robodoc.at, Perl, Jfdwolff, Zizonus, Alexf, Onco p53, OverlordQ, Anythingyouwant, Tail, Atemperman, Rich Farmbrough, Vsmith, Silence, Bender235, Violetriga, Nabla, Sharkford, Shanes, Bobo192, Smalljim, Robhu, Gingko, Nicke Lilltroll~enwiki, Arcadian, Giraffedata, Hooperbloob, Abstraktn, Alansohn, Anthony Appleyard, Ricky81682, Bantman, ClockworkSoul, RainbowOfLight, Woohookitty, Thorpe, Graham87, Rjwilmsi, The wub, Flavr-Savr, FlaBot, RexNL, Narvalo, Jaraalbe, DVdm, Bgwhite, YurikBot, Vagodin, Eraserhead1, Tznkai, Grafen, Arichnad, Lepidoptera, Jaufrec, Epipelagic, Bota47, Leptictidium, CWenger, SmackBot, Jfurr1981, Alksub, EncycloPetey, Bragador, BiT, MalafayaBot, Domthedude001, Милан Јелисавчић, Dreadstar, Drphilharmonic, Ugur Basak Bot~enwiki, SashatoBot, ArglebargleIV, Rory096, Potosino, Sir Nicholas de Mimsy-Porpington, JoseREMY, Kleinburgerei, Mr. Lefty, Inoesomestuff, Manifestation, MTSbot~enwiki, Wilbiddle42, Courcelles, Woodshed, Tawkerbot2, Banedon, SEJohnston, WeggeBot, Treybien, MC10, Gogo Dodo, Chasingsol, Smeazel, Oliver202, West Brom 4ever, James086, Cyclonenim, Luna Santin, Seaphoto, TimVickers, Wayiran, JAnDbot, MER-C, OckRaz, LittleOldMe, VoABot II, JamesBWatson, WhatamIdoing, Allstarecho, Gwern, Jimthompson~enwiki, MartinBot, Anaxial, CommonsDelinker, Nono64, Tgeairn, Svetovid, Zezima8282, Eskimospy, Katalaveno, Dr d12, Jasonasosa, StrayGoose, Bobianite, Leemyster, Bcnof, Lights, VolkovBot, SERSeanCrane, Jeff G., Philip Trueman, Tameeria, Qxz, Blarvink, Madhero88, Burntsauce, Logan, NHRHS2010, Fcady2007, Midjungards, Fanatix, Bfpage, Earthelemental99, Xenophon777, Yerpo, Judicatus, Tombomp, WacoJacko, Onopearls, Hamiltondaniel, Pam519, TheCatalyst31, Atif.t2, ClueBot, Vladkornea, Psypherium, Moguls, Ktr101, Sean Steel, Leonard^Bloom, The Founders Intent, Peter.C, Thingg, 7, Versus22, Johnuniq, Novjunulo, Life of Riley, Lumenos, Jovianeye, Rror, Painking, Avoided, WikHead, NellieBly, Addbot, Brumski, Ronhjones, CactusWriter, Download, Glane23, Omnipedian, Favonian, 5 albert square, Dangles1989, Tide rolls, Bushyballz, Anxietycello, Zorrobot, WikiDreamer Bot, Jarble, Angrysockhop, Legobot, Luckas-bot, Vedran12, Yobot, Jacobs, Maxí, AnakngAraw, JackieBot, Materialscientist, Quebec99, Xqbot, Jayarathina, Saffile, Capricorn42, Spotfixer, Almabot, RibotBOT, Kirin13, KenByers5, Doulos Christos, Thehelpfulbot, Stockprice1977, Prari, Eisengel, TruthIIPower, Doctorwhofan328, Pinethicket, CANTFLAME, Steve2011, Jauhienij, Kalaiarasy, Lotje, Pexego, Ranga e, Purple garden gnome, FoxLogick, Salvio giuliano, Tommy2010, K6ka, Wikmfi, E-citizen, Monterey Bay, Mayur, Donner60, Benjaminalberto, ClueBot NG, This lousy T-shirt, Vacation9, Frietjes, Dictabeard, Zynwyx, Mesoderm, Rezabot, Widr, Meepdeedoo, Titodutta, MusikAnimal, Fontea, GoShow, Bear h, JYBot, Makecat-bot, Sidsandyy, Lugia2453, The Anonymouse, Clucaj, Tempuser00, Padraig Singal, Vuagunny2608, Dheer chudasama, Hubbard1231, Miakirsty123 and Anonymous: 246

- **Cleavage (embryo)** *Source:* https://en.wikipedia.org/wiki/Cleavage_(embryo)?oldid=712060158 *Contributors:* JWSchmidt, Arkuat, Arcadian, Arthena, RoySmith, Phi beta, Jackhynes, Polyparadigm, Joygerhardt, Rjwilmsi, Nihiltres, Flowerparty, YurikBot, Chris Capoccia, Snek01, Daniel Mietchen, 2over0, SmackBot, Zephyris, Yamaguchi⁇⁇, IlliniWikipedian, Shalom Yechiel, Drphilharmonic, Ligulembot, Shinryuu, Alexei Kouprianov, Neelix, Cydebot, Eubanks718, Odmrob, Lauranrg, CommonsDelinker, Boghog, Mikael Häggström, SuW, VolkovBot, Albval, Neparis, Bfpage, SieBot, Nursenjo, ClueBot, Excirial, Alexbot, SkyMaja, BOTarate, PotentialDanger, Koumz, Addbot, Metsavend, Tgm8, Favonian, Anxietycello, Yobot, KamikazeBot, AnomieBOT, Citation bot, Xqbot, Smallman12q, FrescoBot, C.orosco, FloriOn, ChuispastonBot, ClueBot NG, Mesoderm, Helpful Pixie Bot, X -robot- X, Dexbot, Mogism, Me, Myself, and I are Here, Epicgenius, Zorahia, NJSfour and Anonymous: 55

- **Blastomere** *Source:* https://en.wikipedia.org/wiki/Blastomere?oldid=655455893 *Contributors:* AxelBoldt, Lexor, Henrygb, Arcadian, Arthena, SemperBlotto, MarcoTolo, FlaBot, Roboto de Ajvol, EricCHill, Reedy, Jfurr1981, SashatoBot, Novangelis, Rosskey711, JAnDbot, A4bot, ClueBot, Vojtěch Dostál, Addbot, ⁇⁇, Adeliine, Frietjes, Rbj2012, Manningtg, Vicktory7 and Anonymous: 17

- **Morula** *Source:* https://en.wikipedia.org/wiki/Morula?oldid=698483176 *Contributors:* Heron, Hephaestos, DennisDaniels, Lexor, Robbot, Roy-Boy, Bobo192, Arcadian, SpeedyGonsales, Wouterstomp, Metju~enwiki, Benbest, Eyu100, FlaBot, Margosbot~enwiki, YurikBot, Mfero, Lepidoptera, Nolanus, Kubra, SmackBot, InvictaHOG, Jfurr1981, EncycloPetey, Audriusa, Ravi12346, Lottamiata, Robotsintrouble, Neelix, Peter morrell, Daniel, Escarbot, The prophet wizard of the crayon cake, W7347, JAnDbot, SilentWings, Ubiquita, CommonsDelinker, AlphaEta, Sollosonic, VolkovBot, Synthebot, Bfpage, Chhandama, Yerpo, ClueBot, Alexbot, Addbot, CarsracBot, SpBot, Anxietycello, Zorrobot, Fryed-peach, Luckas-bot, Yobot, AnomieBOT, Materialscientist, Maxis ftw, DynamoDegsy, Erud, GrouchoBot, Erik9bot, DrilBot, 3BBOOD, RjwilmsiBot, Amerias, ZéroBot, ClueBot NG, Frietjes, Mesoderm, Helpful Pixie Bot, Snow Blizzard, Iztwoz, Monkbot, Shibbolethink, Tilifa Ocaufa and Anonymous: 41

- **Blastocoel** *Source:* https://en.wikipedia.org/wiki/Blastocoel?oldid=706116708 *Contributors:* Lexor, Merovingian, Onco p53, Arcadian, Seans Potato Business, FlaBot, Margosbot~enwiki, Modify, Kubra, Jonathan.s.kt, SmackBot, BirdValiant, Cryptex, JLCA, Tawkerbot2, Dr. Blofeld, Philg88, CommonsDelinker, NewEnglandYankee, VolkovBot, Vojtěch Dostál, Addbot, Tide rolls, Yobot, AnomieBOT, Gigemag76, Trappist the monk, Klbrain, ZéroBot, Frietjes, Gurt Posh, Makecat-bot, Manníntg, Leprof 7272, Iztwoz, Yahadzija and Anonymous: 21

- **Blastocyst** *Source:* https://en.wikipedia.org/wiki/Blastocyst?oldid=702822738 *Contributors:* AxelBoldt, Magnus Manske, Lexor, Darkwind, Greenrd, Jeffq, Bearcat, Robbot, Cyrius, Giftlite, Robodoc.at, Chowbok, Bender235, Nectarflowed, Arcadian, Giraffedata, Unused000701, Alansohn, Seans Potato Business, Aniketvartak, Tycho, Gimboid13, Graham87, Rjwilmsi, Joffan, FlaBot, Frappyjohn, Tznkai, Wolfmankurd, Chris Capoccia, SpuriousQ, Mfero, Tetsuo, Rmky87, BOT-Superzerocool, Elkman, Werdna, Eaefremov, Ashenai, Ksargent, Sct72, James Mc-Nally, SashatoBot, ArglebargleIV, Epingchris, Sir Nicholas de Mimsy-Porpington, Soulkeeper, Lottamiata, UncleDouggie, Xcentaur, Was a bee, Narayanese, Thijs!bot, Al Lemos, AgentPeppermint, AntiVandalBot, Opelio, JAnDbot, ZDrache, Wikipodium, JaGa, Timothy Titus, J.delanoy, Mikael Häggström, M-le-mot-dit, Mufka, Jonas094, Synthebot, OnlyWayne, SieBot, Milnivri, VVVBot, Heatring, Sunrise, Avenged Eightfold, The Thing That Should Not Be, Mild Bill Hiccup, -Midorihana-, Flatjosh, Novjunulo, Rungladwin, Vojtěch Dostál, Addbot, Diptanshu.D, Chzz, OlEnglish, Legobot, Luckas-bot, Yobot, Fraggle81, Takanjack, AnomieBOT, IRP, Citation bot, Capricorn42, GrouchoBot, Omnipaedista, Listerineman, FrescoBot, Strawbaby, Pknkly, Pinethicket, Kugalskaper, Trappist the monk, Some Wiki Editor, Bohemian89, EmausBot, Jcbsdhbc777, ClueBot NG, Frietjes, Save me, Barry!, Mesoderm, Mohamed CJ, AvocatoBot, Harimiao, Makecat-bot, JakobSteenberg, Manníntg, I am One of Many, Iztwoz, Wildcator, NJSfour, Vicktory7, Singleembryotransfer, Amortias, Shibbolethink, KasparBot, Charleselionel and Anonymous: 116

- **Blastula** *Source:* https://en.wikipedia.org/wiki/Blastula?oldid=708940132 *Contributors:* Bryan Derksen, Lexor, Angela, Bender235, Nectarflowed, Arcadian, Giraffedata, Jumbuck, Alansohn, Axl, Tycho, Metju~enwiki, Drbreznjev, Woohookitty, Mindmatrix, Rjwilmsi, Uwe Gille, FlaBot, Margosbot~enwiki, Dj Capricorn, YurikBot, Wavelength, Shawn81, Ugur Basak, Lepidoptera, Caroline Sanford, InvictaHOG, EncycloPetey, Audriusa, Dl2000, Lottamiata, Bobamnertiopsis, Woodshed, JForget, Thijs!bot, Headbomb, Beelaj., JAnDbot, Leuko, Struthious Bandersnatch, Adavidb, RTBoyce, Silkwilk5, Squids and Chips, TXiKiBoT, Bfpage, Stfg, ClueBot, Mild Bill Hiccup, Puchiko, Excirial, Vojtěch Dostál, Addbot, Ronhjones, Fryed-peach, Luckas-bot, Yobot, AnomieBOT, Xqbot, RibotBOT, Smallman12q, Saiarcot895, FoxBot, RjwilmsiBot, Bohemian89, RockMagnetist, ClueBot NG, Frietjes, Chingla, Mesoderm, Supadude54, Helpful Pixie Bot, Strike Eagle, BattyBot, JYBot, Sminthopsis84, Makecat-bot, Manníntg, Bgumbardo, Dickhitch, Tadala.jumbe, Monkbot, Prakhar vashist, Musashi San17 and Anonymous: 63

- **Inner cell mass** *Source:* https://en.wikipedia.org/wiki/Inner_cell_mass?oldid=710571697 *Contributors:* Pabloes, Arcadian, Seans Potato Business, Woohookitty, BillC, Yamamoto Ichiro, Roboto de Ajvol, Mikalra, Wolfmankurd, Magn0lia, Bluebot, James McNally, Mgiganteus1, Novangelis, Dl2000, Lottamiata, George100, Amalas, Robotsintrouble, Phl3djo, Was a bee, Scarface., GAThrawn22, Beelaj., JAnDbot, Sepul^, Magioladitis, MartinBot, Petter Bøckman, J.delanoy, Mikael Häggström, Iru9k, Sfbergo, Adriennne, PixelBot, Addbot, Yobot, Arthur-Bot, FrescoBot, Rthistle, ZéroBot, MacDaid, Frietjes, Helpful Pixie Bot, KLBot2, Vokesk, Bjorklund21, Lugia2453, SteenthIWbot, Abdullah123456789012345678901234567890, Arielrinon and Anonymous: 15

- **Bilaminar blastocyst** *Source:* https://en.wikipedia.org/wiki/Bilaminar_blastocyst?oldid=706825132 *Contributors:* Arcadian, Giraffedata, Elonka, Maratanos, Kupirijo, MiltonT, CommonsDelinker, Vevebe, ImageRemovalBot, DumZiBoT, Dthomsen8, Addbot, Yobot, John of Reading, ClueBot NG, Frietjes, Jmeyerman8, Cbpatri, Shuvaban Dey and Anonymous: 8

- **Hypoblast** *Source:* https://en.wikipedia.org/wiki/Hypoblast?oldid=657410984 *Contributors:* GreenReaper, Arcadian, Wouterstomp, FlaBot, SmackBot, Delldot, CommonsDelinker, Smirkster, Mikael Häggström, DumZiBoT, Addbot, Luckas-bot, Yobot, Ganímedes, Rrigg021, ZéroBot, Frietjes, Helpful Pixie Bot, BG19bot, LT910001, Allielew00, YueM and Anonymous: 5

- **Epiblast** *Source:* https://en.wikipedia.org/wiki/Epiblast?oldid=703983093 *Contributors:* Saucepan, Arcadian, Benbest, FlaBot, Mikalra, Groogle, Notch, Sbmehta, Robotsintrouble, Kupirijo, .anacondabot, KylieTastic, Fimbriata, Adriennne, Blahman1985, Addbot, LaaknorBot, Agl5008, Dangles1989, Yobot, Rubinbot, Xqbot, ZéroBot, Frietjes, Iztwoz, Allielew00, JeffreyDTse and Anonymous: 6

- **Trilaminar blastocyst** *Source:* https://en.wikipedia.org/wiki/Trilaminar_blastocyst?oldid=666120564 *Contributors:* Arcadian, Elkman, Delldot, Serephine, Alphachimpbot, SporkBot, Frietjes, KLBot2, Vokesk and Hamoudafg

- **Germ layer** *Source:* https://en.wikipedia.org/wiki/Germ_layer?oldid=664575196 *Contributors:* Heron, Lexor, Jebba, Nikai, Zarius, RodC, Dave6, Curps, PDH, Fungus Guy, Nina Gerlach, Tinus, Bender235, CheekyMonkey, RoyBoy, Arcadian, Jag123, Alansohn, Wouterstomp, Tycho, Ceyockey, Woohookitty, Benbest, Rjwilmsi, Shao, Dj Capricorn, Nephron, Zwobot, Wknight94, RupertMillard, KnowledgeOfSelf, Kipmaster, Gilliam, J.Steinbock, Bluebot, Tamfang, Radagast83, TheLimbicOne, Bansp, Clicketyclack, Werlop, Epingchris, Jon186, Lottamiata, Vsoulremix, Jamoche, A876, Thijs!bot, Kilva, GAThrawn22, AntiVandalBot, Alphachimpbot, JAnDbot, RuthieK, STBot, Wlodzimierz, CFCF, Ginsengbomb, Jotunn, JBarno, VolkovBot, LeilaniLad, TXiKiBoT, BotKung, AlleborgoBot, SieBot, Kochipoik, Anchor Link Bot, ClueBot, Eric Van Bogaert, Addbot, 2enable, Yobot, AnomieBOT, Lapabc, RibotBOT, FrescoBot, D'ohBot, Fama Clamosa, TjBot, JaysonSunshine, EmausBot, GoingBatty, Donner60, ClueBot NG, Harps21, Mesoderm, Newyorkadam, BG19bot, Biolprof, Kenneth.jh.han, Manníntg, Biologize, Iztwoz, Sonicnation, Comp.arch, Mendoza.m420 and Anonymous: 76

- **Archenteron** *Source:* https://en.wikipedia.org/wiki/Archenteron?oldid=708769248 *Contributors:* SimonP, Tagishsimon, Arcadian, RJFJR, 2004-12-29T22:45Z, Jade05, Dysepsion, Dbutler1986, Quale, FlaBot, YurikBot, Gaius Cornelius, EncycloPetey, James McNally, Tamarkot, Kupirijo, Thijs!bot, Escarbot, VolkovBot, Alexbot, Vojtěch Dostál, Paddy Simcox, Addbot, Rhodospirillum, Luckas-bot, Materialscientist, Jhzjas, Jonnjohn8292, Emilyharry, HRoestBot, Mesoderm, Iztwoz and Anonymous: 23

- **Primitive streak** *Source:* https://en.wikipedia.org/wiki/Primitive_streak?oldid=702589010 *Contributors:* William Avery, GreenReaper, Arcadian, Woohookitty, Rjwilmsi, Uwe Gille, Hairy Dude, Takwish, George Church, Twr57, Bejnar, Alexey Feldgendler, Robotsintrouble, Kupirijo, GAThrawn22, JamesBWatson, Eltiburon, Una Smith, PipepBot, Callinus, DumZiBoT, Addbot, Tassedethe, Verazzano, Yobot, Citation bot,

FrescoBot, Citation bot 1, DrilBot, RedBot, A p3rson, RjwilmsiBot, Doughorner, Afiguero, ZéroBot, Hon-3s-T, Frietjes, Hazhk, Mesoderm, Helpful Pixie Bot, BG19bot, Nelg, Dickhitch, CensoredScribe, NJSfour, Bullets and Bracelets and Anonymous: 14

- **Primitive pit** *Source:* https://en.wikipedia.org/wiki/Primitive_pit?oldid=657441969 *Contributors:* Arcadian, Was a bee, Katharineamy, Yobot, Frietjes, Iztwoz and Anonymous: 1

- **Primitive knot** *Source:* https://en.wikipedia.org/wiki/Primitive_knot?oldid=702588991 *Contributors:* Magnus Manske, Sunray, Nmg20, Arcadian, Melaen, Rjwilmsi, Gaius Cornelius, LaurenCole, TheLimbicOne, Cydebot, Kupirijo, Addbot, DOI bot, LaaknorBot, Yobot, Frietjes, Blairwal, Amdou4 and Anonymous: 9

- **Gastrulation** *Source:* https://en.wikipedia.org/wiki/Gastrulation?oldid=712374372 *Contributors:* The Anome, Malcolm Farmer, Lexor, Angela, MichaK, Mackensen, Robbot, Dina, Robodoc.at, JeffreyN, Discospinster, Arcadian, SpeedyGonsales, MPerel, Snowolf, KingTT, Tycho, Metju~enwiki, Alai, Ceyockey, Woohookitty, Hendrik Fuß, BD2412, Kissekatt, Rjwilmsi, Agrumer, FlaBot, Jrtayloriv, Banaticus, YurikBot, Wavelength, SLATE, GeeJo, Zwobot, BOT-Superzerocool, Drosboro, Mike Dillon, SmackBot, Saravask, Essent, Radagast83, TheLimbicOne, T-borg, James McNally, Clicketyclack, Mgiganteus1, Mike Doughney, Vanished user, Lottamiata, Zinzen, TheTito, Kupirijo, Eubanks718, Odmrob, GAThrawn22, Nipisiquit, JAnDbot, Rothorpe, Bissinger, R'n'B, Wlodzimierz, Mikael Häggström, Dexter prog, Nwbeeson, Squids and Chips, VolkovBot, Philip Trueman, Etruria, Earthdirt, Mwilso24, Ian Glenn, Happysailor, Flyer22 Reborn, Hxhbot, SimonTrew, The Thing That Should Not Be, Robomanx, Ndenison, Iandiver, Msgarrett, Excirial, Alexbot, Flatjosh, Wnt, Alboyle, Frostus, Addbot, Tide rolls, Cesiumfrog, Fryed-peach, Rhodospirillum, Yobot, Vini 17bot5, AnomieBOT, Kyng, TheChymera, FrescoBot, Sinekonata, RandomStringOfCharacters, FoxBot, Trappist the monk, Makki98, Agent Smith (The Matrix), John of Reading, MirekDve, GoingBatty, Hardrockcrossing, Palomitaviajera, White Trillium, Matthewrbowker, ClueBot NG, Mesoderm, Temperamental1, Widr, Helpful Pixie Bot, B2322858, Dean72, Scanbre, CeraBot, Biolprof, Khazar2, Apynekeeper, Boydstra, Iztwoz, Haned6011, Depalmal, LT910001, NJSfour, T1d7, CV9933, JeremiahY and Anonymous: 78

- **Primitive groove** *Source:* https://en.wikipedia.org/wiki/Primitive_groove?oldid=657014453 *Contributors:* Arcadian, Gyre, Elkman, Caerwine, SmackBot, Gogo Dodo, Was a bee, Yobot, Frietjes and Anonymous: 2

- **Regional differentiation** *Source:* https://en.wikipedia.org/wiki/Regional_differentiation?oldid=707989059 *Contributors:* BenKovitz, Gary, Rjwilmsi, Nihiltres, GeeJo, Melchoir, Kazkaskazkasako, Celefin, Michael Bednarek, Eubanks718, Alaibot, Nono64, Mikael Häggström, Flyer22 Reborn, Forluvoft, ClueBot, DOI bot, Yobot, FrescoBot, El Mayimbe, John of Reading, Nyxhadanielle, Duester, Helpful Pixie Bot, BG19bot, Rob Hurt, Yahadzija, Monkbot, JMWSlack and Anonymous: 20

- **Embryonic disc** *Source:* https://en.wikipedia.org/wiki/Embryonic_disc?oldid=657015080 *Contributors:* Arcadian, Rmky87, Elkman, Martijn Hoekstra, Was a bee, Dig deeper, Sun Creator, SilvonenBot, Addbot, Yobot, Frietjes and FamAD123

- **Ectoderm** *Source:* https://en.wikipedia.org/wiki/Ectoderm?oldid=704650435 *Contributors:* Alex.tan, Ellywa, Docu, Altenmann, Fuelbottle, Thijs!, Woggly, Sayeth, Nina Gerlach, JemeL, Bender235, Smalljim, AllyUnion, Arcadian, Jag123, Wouterstomp, Wtmitchell, Tycho, Hsmith254, Julien Tuerlinckx, Ruziklan, BD2412, Rjwilmsi, Margosbot~enwiki, Shao, YurikBot, Nicke L, Chodges, Chazz88, J.Steinbock, Radagast83, TheLimbicOne, Clicketyclack, SashatoBot, Epingchris, Robotsintrouble, Hebrides, Thijs!bot, GAThrawn22, Bbman4ever, Smartse, Alphachimpbot, JAnDbot, Txomin, RebelRobot, Kelleyo2l, Nyq, Japo, The cattr, Rettetast, Verdatum, Boghog, Chiswick Chap, Squids and Chips, VolkovBot, AlnoktaBOT, Drgarden, PipepBot, Niceguyedc, Profipix, Wikalliz, Addbot, Kongr43gpen, Guffydrawers, Verazzano, Luckas-bot, Yobot, Lynntyler, JackieBot, Citation bot, Smallman12q, Erik9bot, FrescoBot, Pinethicket, Verrr55449988776655, FoxBot, RjwilmsiBot, EmausBot, WikitanvirBot, GoingBatty, Daonguyen95, Fæ, Ws04, Froggy980, ClueBot NG, Harps21, Qbobdole, Mesoderm, Hakeleh, Rytyho usa, BattyBot, Khazar2, Dexbot, Kenneth.jh.han, Tyler7810, Olefsky, Aadharm, Hegarty.michael.c, Monkbot, Alawy88, CAPTAIN RAJU and Anonymous: 46

- **Surface ectoderm** *Source:* https://en.wikipedia.org/wiki/Surface_ectoderm?oldid=657028633 *Contributors:* Xezbeth, Arcadian, A876, Was a bee, Nono64, Yobot, EmausBot, ZéroBot, Feral mage, Frietjes, Mmhrmhrm, Fgegypt and Anonymous: 1

- **Neuroectoderm** *Source:* https://en.wikipedia.org/wiki/Neuroectoderm?oldid=666119566 *Contributors:* Xezbeth, Arcadian, Uwe Gille, Bhny, Was a bee, Synthebot, Alexbot, Vojtěch Dostál, Addbot, Yobot, Beeswaxcandle, Borsukers, SporkBot, Cjwhitetds767, Frietjes, Master ecclesias and Anonymous: 9

- **Somatopleuric mesenchyme** *Source:* https://en.wikipedia.org/wiki/Somatopleuric_mesenchyme?oldid=657014948 *Contributors:* CBDunkerson, Tom harrison, Arcadian, FlaBot, Winhunter, Consumed Crustacean, Erachima, Deskana, Elkman, Caerwine, Yanksox, Khoikhoi, Weregerbil, Ryulong, Was a bee, Wildnox, B, TheNameGame, Interrogation, Waterboarder, BreathingApparatus, Bullets&Love, ShoYoAss, MisterPedersen, SaveTheWails, WalesSaveUs, SecondCityBest, EverythingsErased, ForMyFolkers, WhenShesTenFeetTalll, Air Races, Sealing Fan, NCB1148, Xenocentrist, TheChainGang, AnImpropperFraction, Chekew Sir Bloque, BloodClots, Ouedbirdwatcher, Addbot, Luckas-bot, Yobot, Frietjes and Anonymous: 3

- **Neurulation** *Source:* https://en.wikipedia.org/wiki/Neurulation?oldid=712203917 *Contributors:* Diberri, Delta G, Arcadian, Tycho, Kurzon, RxS, Rjwilmsi, FlaBot, Gurch, Hitokirishinji, YurikBot, Mr Frosty, Supten, Chefyingi, SmackBot, Radak, EncycloPetey, Eskimbot, J.Steinbock, TheLimbicOne, Shushruth, Confuseddave, Mgiganteus1, ChrisCork, Gareth1337, Eubanks718, GAThrawn22, Magioladitis, Stephenchou0722, Crumja, Nono64, CFCF, Mikael Häggström, VolkovBot, PNG crusade bot, Loki~enwiki, Poltair, Srikandicinta, Anchor Link Bot, Addbot, Luckas-bot, Yobot, AnomieBOT, Rubinbot, JohnnyA54, Ahambhavami, Nirmos, Linguisticgeek, RjwilmsiBot, Terrasigillata, ZéroBot, Italienmoose, Whoop whoop pull up, Frietjes, Mesoderm, Helpful Pixie Bot, BG19bot, Bush6984, BattyBot, ChrisGualtieri, Jlesk, Iztwoz, N1424, Therazzz, Monkbot, Sarr Cat, Ssmith3284, Jbull991 and Anonymous: 42

- **Neural crest** *Source:* https://en.wikipedia.org/wiki/Neural_crest?oldid=688932144 *Contributors:* The Anome, Diberri, Binadot, Arcadian, Nik42, Myrtlegroggins, Mandarax, Rjwilmsi, Lar, Killdevil, SmackBot, EncycloPetey, Kipmaster, Eug, Armeria, J.Steinbock, Paxfeline, Tim-Bentley, Colonies Chris, TheLimbicOne, Drphilharmonic, Takowl, Bejnar, Nishkid64, Werlop, Dr.saptarshi, Novangelis, Robotsintrouble, Ruslik0, Eubanks718, Alaibot, Thijs!bot, CopperKettle, Kauczuk, GetAgrippa, Gwern, NikNaks, Mikael Häggström, TXiKiBoT, Luuva, AS, Bgordski, Mcvmccarty, Estevoaei, Agentareas, BOTarate, DumZiBoT, Vojtěch Dostál, Shunju-kun, Addbot, DOI bot, Skamnelis, SpBot, Tahmmo, Lightbot, Verazzano, مانی, Yobot, LilHelpa, Tylersquare, FrescoBot, Goldstein.ron, Shiningheart, Acatyes, Ibderm, Abitua, Rob.weitemeyer, Feral mage, Povilasq, Ostreet, Wikitavanti, Etche homo, KMaher123, ChrisGualtieri, Dexbot, Therazzz, Comp.arch, Arielrinon, Tilifa Ocaufa and Anonymous: 43

- **Endoderm** *Source:* https://en.wikipedia.org/wiki/Endoderm?oldid=711009687 *Contributors:* Alex.tan, Lexor, Docu, MichaK, Topbanana, Mike Rosoft, Nina Gerlach, Bobo192, Arcadian, HasharBot~enwiki, Hovea, Rjwilmsi, Ligulem, FlaBot, Shao, YurikBot, Postglock, Hede2000, Dysmorodrepanis~enwiki, Zwobot, DRosenbach, KnightRider~enwiki, SmackBot, J.Steinbock, Tsca.bot, TheLimbicOne, Clicketyclack, Sashato-Bot, Was a bee, Narayanese, Thijs!bot, Barticus88, GAThrawn22, Oreo Priest, KineticScientist, Exiledone, DrewBY, J.delanoy, Lionesschan, VolkovBot, TXiKiBoT, Una Smith, Drgarden, ClueBot, NuclearWarfare, Addbot, Luckas-bot, Yobot, Amirobot, Rubinbot, Xqbot, Smallman12q, TobeBot, EmausBot, AvicBot, Qbobdole, Frietjes, Mesoderm, Widr, Reza luke, Webclient101, Kenneth.jh.han, Daniilkaz, Richjoo, Monkbot, Qzd, Myoglobin and Anonymous: 39

- **Splanchnopleuric mesenchyme** *Source:* https://en.wikipedia.org/wiki/Splanchnopleuric_mesenchyme?oldid=712081683 *Contributors:* CB-Dunkerson, Tom harrison, Arcadian, Benbest, FlaBot, Winhunter, Consumed Crustacean, Erachima, Deskana, Elkman, Caerwine, Yanksox, Khoikhoi, Ryulong, Was a bee, Wildnox, Jmauck, TheNameGame, Interrogation, Waterboarder, BreathingApparatus, Bullets&Love, ShoYoAss, MisterPedersen, SaveTheWails, WalesSaveUs, SecondCityBest, EverythingsErased, ForMyFolkers, Air Races, Sealing Fan, NCB1148, Xenocentrist, TheChainGang, AnImpropperFraction, Chekew Sir Bloque, BloodClots, Jarrodtrainque, Addbot, Yobot, Reluwa, Frietjes and Anonymous: 6

- **Mesoderm** *Source:* https://en.wikipedia.org/wiki/Mesoderm?oldid=709523944 *Contributors:* Magnus Manske, Alex.tan, Lexor, Kimiko, Emperorbma, Fuelbottle, Diberri, Creidieki, T-Boy, HCA, CanisRufus, Arcadian, Jag123, Tycho, Georgia guy, Mandarax, Zxccxz, FlaBot, Shao, Bgwhite, YurikBot, NTBot~enwiki, DRosenbach, Closedmouth, J.Steinbock, Khoikhoi, Radagast83, TheLimbicOne, Vina-iwbot~enwiki, Clicketyclack, FrozenMan, Mgiganteus1, RelentlessRecusant, Thijs!bot, Peter morrell, GAThrawn22, Alphachimpbot, Gökhan, Txomin, Magioladitis, AlphaEta, Philcha, Numbo3, Nbauman, AlnoktaBOT, Oshwah, Broadbot, AlleborgoBot, SieBot, Loquetudigas, PixelBot, Dthomsen8, Addbot, Download, ماني, Luckas-bot, Yobot, Atgnclk, AnomieBOT, JackieBot, Citation bot, Xqbot, Macholl, Sviolante, Machn, DrilBot, Hamtechperson, Wdanbae, EmausBot, Dewritech, Dcirovic, Werieth, ClueBot NG, Harps21, Cwmhiraeth, Mesoderm, Helpful Pixie Bot, Reza luke, Hakeleh, Smettems, Kenneth.jh.han, JakobSteenberg, N1424, DrLinguini, Depalmal, SachiPaulete, Alejandra Sotres, Monkbot, Arielrinon, Ritoxavi, Editor N and Anonymous: 53

- **Axial mesoderm** *Source:* https://en.wikipedia.org/wiki/Axial_mesoderm?oldid=689661170 *Contributors:* Diberri, Arcadian, Axyjo, Unused000701, Elkman, Caerwine, Was a bee, GAThrawn22, Cmungall, Yobot, DrilBot, EmausBot, Whoop whoop pull up, Frietjes, Iztwoz, Fgegypt, Equinox and Anonymous: 2

- **Paraxial mesoderm** *Source:* https://en.wikipedia.org/wiki/Paraxial_mesoderm?oldid=691497778 *Contributors:* Diberri, Arcadian, Unused000701, Marudubshinki, BD2412, BOT-Superzerocool, Elkman, Caerwine, Novangelis, Robotsintrouble, Meno25, Was a bee, Addbot, Yobot, John of Reading, ZéroBot, Frietjes, Helpful Pixie Bot, BG19bot, Dexbot, Eprotasiuk, Iztwoz, Arielrinon, Legenderfox and Anonymous: 4

- **Somite** *Source:* https://en.wikipedia.org/wiki/Somite?oldid=712472280 *Contributors:* The Anome, Edward, Lexor, Diberri, Delta G, Discospinster, Bobo192, Arcadian, RJFJR, GregorB, MarcoTolo, BD2412, Rjwilmsi, Gadig, RexNL, Bgwhite, YurikBot, Voidxor, DRosenbach, Avalon, Calvin08, Lycaon, SmackBot, Agibso02, EncycloPetey, Kostmo, TheLimbicOne, Henning Makholm, Confuseddave, Aleator, Mgiganteus1, Novangelis, Thijs!bot, Bjordan555, Lauranrg, Nick Number, Oreo Priest, Nyttend, R'n'B, Natsirtguy, CFCF, Colincbn, WarthogDemon, Chiswick Chap, Nwbeeson, Chrispruet, LeadSongDog, Mild Bill Hiccup, Frostus, MystBot, Addbot, Angrense, MuZemike, Yobot, AnomieBOT, Lynntyler, EmausBot, ZéroBot, Koala0090, Gongoozler123, ClueBot NG, Michaelcshapiro, Frietjes, Widr, BG19bot, IluvatarBot, BattyBot, Amulshah7, Biolprof, KellyMurray, Dexbot, Eprotasiuk, Iztwoz, LT910001, Monkbot, Arielrinon and Anonymous: 35

- **Somitomere** *Source:* https://en.wikipedia.org/wiki/Somitomere?oldid=635441708 *Contributors:* Lexor, Diberri, Arcadian, Slambo, RJFJR, The JPS, Graham87, EncycloPetey, TheLimbicOne, Euchiasmus, MECU, Yobot, LilHelpa, Frietjes, Helpful Pixie Bot, Iztwoz and Anonymous: 7

- **Intermediate mesoderm** *Source:* https://en.wikipedia.org/wiki/Intermediate_mesoderm?oldid=691539043 *Contributors:* Diberri, Timpo, Arcadian, Unused000701, Elkman, Caerwine, SmackBot, Patho~enwiki, CmdrObot, Was a bee, GAThrawn22, Flyer22 Reborn, Yobot, Anypodetos, DrilBot, ClueBot NG, Mrslisa, Frietjes, Jmeyerman8, TylerDurden8823, I am Trinity and Anonymous: 6

- **Lateral plate mesoderm** *Source:* https://en.wikipedia.org/wiki/Lateral_plate_mesoderm?oldid=703288056 *Contributors:* Diberri, Arcadian, Unused000701, Elkman, Caerwine, Novangelis, Was a bee, PKT, GAThrawn22, Exiledone, Michaeldsuarez, Icarusgeek, DOI bot, Yobot, Cnk5hoya, DrilBot, Frietjes, Helpful Pixie Bot, WikiHannibal, JakobSteenberg, Iztwoz, Arielrinon and Anonymous: 8

- **Intra-embryonic coelom** *Source:* https://en.wikipedia.org/wiki/Intra-embryonic_coelom?oldid=686674403 *Contributors:* Bearcat, Arcadian, Rjwilmsi, Malcolma, LonghornDIT, Jbeans, Yobot, Frietjes, KLBot2, June Bea, Fgegypt and Anonymous: 2

40.2.2 Images

- **File:2037_Embryonic_Development_of_Heart.jpg** *Source:* https://upload.wikimedia.org/wikipedia/commons/7/74/2037_Embryonic_Development_of_Heart.jpg *License:* CC BY 3.0 *Contributors:* Anatomy & Physiology, Connexions Web site. http://cnx.org/content/col11496/1.6/, Jun 19, 2013. *Original artist:* OpenStax College

- **File:2910_The_Placenta-02.jpg** *Source:* https://upload.wikimedia.org/wikipedia/commons/2/2d/2910_The_Placenta-02.jpg *License:* CC BY 3.0 *Contributors:* Anatomy & Physiology, Connexions Web site. http://cnx.org/content/col11496/1.6/, Jun 19, 2013. *Original artist:* OpenStax College

- **File:2912_Neurulation-02.jpg** *Source:* https://upload.wikimedia.org/wikipedia/commons/0/0f/2912_Neurulation-02.jpg *License:* CC BY 3.0 *Contributors:* Anatomy & Physiology, Connexions Web site. http://cnx.org/content/col11496/1.6/, Jun 19, 2013. *Original artist:* OpenStax College

- **File:Acrosome_reaction_diagram_en.svg** *Source:* https://upload.wikimedia.org/wikipedia/commons/8/83/Acrosome_reaction_diagram_en.svg *License:* Public domain *Contributors:* I made this diagram myself using the diagrams on: [1] , [2],and [3]. I used Adobe Illustrator to do it including pgf data, and I am freeing it into public domain. LadyofHats. *Original artist:* LadyofHats.

- **File:Anatomy_of_an_egg_unlabeled_horizontal.svg** *Source:*   https://upload.wikimedia.org/wikipedia/commons/b/b3/Anatomy_of_an_ egg_unlabeled_horizontal.svg *License:* CC-BY-SA-3.0 *Contributors:* graphic created by de:Benutzer:Horst Frank, SVG version by cs:User:-xfi- *Original artist:* de:Benutzer:Horst Frank, SVG code cs:User:-xfi-, rotation and text removal by User:Kjoonlee

- **File:Blastocyst_English.svg** *Source:* https://upload.wikimedia.org/wikipedia/commons/7/72/Blastocyst_English.svg *License:* CC-BY-SA-3.0 *Contributors:* Blastocyst.png *Original artist:* Seans Potato Business (derivative of the source cited above)

- **File:Blastula_(PSF).jpg** *Source:* https://upload.wikimedia.org/wikipedia/commons/1/1e/Blastula_%28PSF%29.jpg *License:* Public domain *Contributors:* Archives of Pearson Scott Foresman, donated to the Wikimedia Foundation *Original artist:* Pearson Scott Foresman

- **File:Blausen_0404_Fertilization.png** *Source:* https://upload.wikimedia.org/wikipedia/commons/7/79/Blausen_0404_Fertilization.png *License:* CC BY 3.0 *Contributors:* Own work *Original artist:* BruceBlaus. When using this image in external sources it can be cited as:

- **File:Bone.png** *Source:* https://upload.wikimedia.org/wikipedia/commons/2/2c/Bone.png *License:* CC BY 3.0 *Contributors:* [1] [2] *Original artist:* BanzaiTokyo

- **File:Cat_brain.jpg** *Source:* https://upload.wikimedia.org/wikipedia/commons/e/e6/Cat_brain.jpg *License:* Public domain *Contributors:* ? *Original artist:* ?

- **File:Commons-logo.svg** *Source:* https://upload.wikimedia.org/wikipedia/en/4/4a/Commons-logo.svg *License:* CC-BY-SA-3.0 *Contributors:* ? *Original artist:* ?

- **File:Development_of_the_neural_tube.png** *Source:* https://upload.wikimedia.org/wikipedia/commons/4/4c/Development_of_the_neural_ tube.png *License:* Public domain *Contributors:* Figure 6 (p. 24) of "The anatomy of the nervous system" by Stephen Walter Ranson, published W.B. Saunders, 1920 *Original artist:* user:Looie496 created file, original artist unknown

- **File:Ectoderm.png** *Source:* https://upload.wikimedia.org/wikipedia/commons/1/1d/Ectoderm.png *License:* CC-BY-SA-3.0 *Contributors:* ? *Original artist:* ?

- **File:EctodermalSpecification.png** *Source:* https://upload.wikimedia.org/wikipedia/commons/4/47/EctodermalSpecification.png *License:* Public domain *Contributors:* Own work *Original artist:* Sofia mr007

- **File:Edit-clear.svg** *Source:* https://upload.wikimedia.org/wikipedia/en/f/f2/Edit-clear.svg *License:* Public domain *Contributors:* The *Tango! Desktop Project. Original artist:*

  The people from the Tango! project. And according to the meta-data in the file, specifically: "Andreas Nilsson, and Jakub Steiner (although minimally)."

- **File:Embryo,_8_cells.jpg** *Source:* https://upload.wikimedia.org/wikipedia/commons/6/6b/Embryo%2C_8_cells.jpg *License:* Public domain *Contributors:* ? *Original artist:* ekem, Courtesy: RWJMS IVF Program

- **File:Endoderm2.png** *Source:* https://upload.wikimedia.org/wikipedia/commons/c/c0/Endoderm2.png *License:* Public domain *Contributors:* http://en.wikipedia.org/wiki/Image:Endoderm2.png *Original artist:* en:User:J.Steinbock

- **File:Epithelial–mesenchymal_transition_scheme.png** *Source:*   https://upload.wikimedia.org/wikipedia/commons/5/54/Epithelial%E2% 80%93mesenchymal_transition_scheme.png *License:* CC BY 3.0 *Contributors:* https://en.wikipedia.org/wiki/File:Screen_shot_2010-04-04_ at_8.34.23_PM.png#file *Original artist:* Hardrockcrossing

- **File:Equal_vs_unequal_cleavage.jpg** *Source:* https://upload.wikimedia.org/wikipedia/commons/6/68/Equal_vs_unequal_cleavage.jpg *License:* Public domain *Contributors:* Transferred from en.wikipedia to Commons by SreeBot. *Original artist:* C.orosco at en.wikipedia

- **File:Formation_of_the_Primitive_Streak.pdf**   *Source:*   https://upload.wikimedia.org/wikipedia/commons/f/f8/Formation_of_the_ Primitive_Streak.pdf *License:* Public domain *Contributors:* Own work *Original artist:* Afiguero

- **File:GRN_Crest.png** *Source:* https://upload.wikimedia.org/wikipedia/en/c/c8/GRN_Crest.png *License:* PD *Contributors:* ? *Original artist:* ?

- **File:Gastrulation.png** *Source:* https://upload.wikimedia.org/wikipedia/commons/3/31/Gastrulation.png *License:* Public domain *Contributors:* Own work *Original artist:* Pidalka44

- **File:Germ_layers.jpg** *Source:* https://upload.wikimedia.org/wikipedia/commons/5/52/Germ_layers.jpg *License:* CC BY-SA 3.0 *Contributors:* http://cnx.org/resources/cdd4d14f0c1cde804e5a84495390806c/Figure_43_05_04.jpg *Original artist:* CNX

- **File:Gray13.png** *Source:* https://upload.wikimedia.org/wikipedia/commons/f/f4/Gray13.png *License:* Public domain *Contributors:* Henry Gray (1918) *Anatomy of the Human Body* (See "Book" section below)

  Original artist: Henry Vandyke Carter

- **File:Gray18.png** *Source:* https://upload.wikimedia.org/wikipedia/commons/9/9e/Gray18.png *License:* Public domain *Contributors:* Henry Gray (1918) *Anatomy of the Human Body* (See "Book" section below)

  Original artist: Henry Vandyke Carter

- **File:Gray36.png** *Source:* https://upload.wikimedia.org/wikipedia/commons/0/0d/Gray36.png *License:* Public domain *Contributors:* Henry Gray (1918) *Anatomy of the Human Body* (See "Book" section below)

  Original artist: Henry Vandyke Carter

- **File:Gray64.png** *Source:* https://upload.wikimedia.org/wikipedia/commons/f/f3/Gray64.png *License:* Public domain *Contributors:* Henry Gray (1918) *Anatomy of the Human Body* (See "Book" section below)

  Original artist: Henry Vandyke Carter

- **File:Gray640.png** *Source:* https://upload.wikimedia.org/wikipedia/commons/2/20/Gray640.png *License:* Public domain *Contributors:* Henry Gray (1918) *Anatomy of the Human Body* (See "Book" section below)

  Original artist: Henry Vandyke Carter

- **File:Gray65.png** *Source:* https://upload.wikimedia.org/wikipedia/commons/9/96/Gray65.png *License:* Public domain *Contributors:* Henry Gray (1918) *Anatomy of the Human Body* (See "Book" section below)
  Original artist: Henry Vandyke Carter

- **File:Gray947.png** *Source:* https://upload.wikimedia.org/wikipedia/commons/3/38/Gray947.png *License:* Public domain *Contributors:* Henry Gray (1918) *Anatomy of the Human Body* (See "Book" section below)
  Original artist: Henry Vandyke Carter

- **File:HED.jpg** *Source:* https://upload.wikimedia.org/wikipedia/commons/f/f6/HED.jpg *License:* CC BY 2.0 *Contributors:* Lisbet K Lind, Christina Stecksén-Blicks, Kristina Lejon, Marcus Schmitt-Egenolf. EDAR mutation in autosomal dominant hypohidrotic ectodermal dysplasia in two Swedish families. BMC Med Genet. 7, 80. 2006. PMID 17125505. *Original artist:* ?

- **File:Human.png** *Source:* https://upload.wikimedia.org/wikipedia/commons/6/6e/Human.png *License:* Public domain *Contributors:* This is a crop (with trivial modifications) of Image:PPlaqueLarge.png, which is itself derived from Image:Pioneer10-plaque.jpg. *Original artist:* NASA

- **File:HumanEmbryogenesis.svg** *Source:* https://upload.wikimedia.org/wikipedia/commons/0/06/HumanEmbryogenesis.svg *License:* CC BY-SA 3.0 *Contributors:* SVG version of <a href='//commons.wikimedia.org/wiki/File:Human_embryogenesis_-2.png' class='image'><img alt='Human embryogenesis —2.png' src='https://upload.wikimedia.org/wikipedia/commons/thumb/6/66/Human_embryogenesis_-2.png/ 100px-Human_embryogenesis_-2.png' width='100' height='80' srcset='https://upload.wikimedia.org/wikipedia/commons/thumb/6/66/ Human_embryogenesis_-2.png/150px-Human_embryogenesis_-2.png 1.5x, https://upload.wikimedia.org/wikipedia/commons/thumb/6/66/ Human_embryogenesis_-2.png/200px-Human_embryogenesis_-2.png 2x' data-file-width='1197' data-file-height='956' /></a>. *Original artist:* Zephyris

- **File:Human_Fertilization.png** *Source:* https://upload.wikimedia.org/wikipedia/commons/5/5b/Human_Fertilization.png *License:* CC BY-SA 3.0 *Contributors:* Own work *Original artist:* Ttrue12

- **File:ICM_signaling.jpg** *Source:* https://upload.wikimedia.org/wikipedia/commons/b/b7/ICM_signaling.jpg *License:* Public domain *Contributors:* Own work *Original artist:* Rthistle

- **File:Mesoderm.png** *Source:* https://upload.wikimedia.org/wikipedia/commons/e/e8/Mesoderm.png *License:* Public domain *Contributors:* ? *Original artist:* J.Steinbock

- **File:Neural_crest.svg** *Source:* https://upload.wikimedia.org/wikipedia/commons/6/60/Neural_crest.svg *License:* CC BY-SA 3.0 *Contributors:* File:Neural_Crest.png *Original artist:* NikNaks

- **File:Neuro_logo.png** *Source:* https://upload.wikimedia.org/wikipedia/commons/f/f8/Neuro_logo.png *License:* Public domain *Contributors:* The PNG crusade bot automatically converted this image to the more efficient PNG format. ; 19:29:54, 18 April 2007 (UTC) Remember the dot (*ShouldBePNG*) *Original artist:* The PNG crusade bot automatically converted this image to the more efficient PNG format. 19:29:54, 18 April 2007 (UTC) Remember the dot (*ShouldBePNG*)

- **File:Nuvola_kdict_glass.svg** *Source:* https://upload.wikimedia.org/wikipedia/commons/1/18/Nuvola_kdict_glass.svg *License:* LGPL *Contributors:*

- Nuvola_apps_kdict.svg *Original artist:* Nuvola_apps_kdict.svg: *Nuvola_apps_kdict.png: user:David_Vignoni

- **File:Protovsdeuterostomes.svg** *Source:* https://upload.wikimedia.org/wikipedia/commons/6/65/Protovsdeuterostomes.svg *License:* CC BY-SA 3.0 *Contributors:* Own work *Original artist:* This vector image was created with Inkscape.

- **File:Question_book-new.svg** *Source:* https://upload.wikimedia.org/wikipedia/en/9/99/Question_book-new.svg *License:* Cc-by-sa-3.0 *Contributors:*
  Created from scratch in Adobe Illustrator. Based on Image:Question book.png created by User:Equazcion *Original artist:* Tkgd2007

- **File:Sobo_1909_621.png** *Source:* https://upload.wikimedia.org/wikipedia/commons/6/6e/Sobo_1909_621.png *License:* Public domain *Contributors:* Atlas and Text-book of Human Anatomy Volume III Vascular System, Lymphatic system, Nervous system and Sense Organs *Original artist:* Dr. Johannes Sobotta

- **File:Spiral_cleavage_in_Trochus.png** *Source:* https://upload.wikimedia.org/wikipedia/commons/2/2d/Spiral_cleavage_in_Trochus.png *License:* CC BY 2.5 *Contributors:* Goulding MQ. 2009. <a data-x-rel='nofollow' class='external text' href='http://www.plosone.org/article/info% 3Adoi%2F10.1371%2Fjournal.pone.0005506'>*Cell Lineage of the Ilyanassa Embryo: Evolutionary Acceleration of Regional Differentiation during Early Development*</a>. PLoS ONE 4(5): e5506. doi:10.1371/journal.pone.0005506 Figure 1 TIFF *Original artist:* Morgan Q. Goulding

- **File:Teratoma_2_low_mag.jpg** *Source:* https://upload.wikimedia.org/wikipedia/commons/1/11/Teratoma_2_low_mag.jpg *License:* CC BY-SA 3.0 *Contributors:* Own work *Original artist:* Nephron

- **File:Translation_to_english_arrow.svg** *Source:* https://upload.wikimedia.org/wikipedia/commons/8/8a/Translation_to_english_arrow.svg *License:* CC-BY-SA-3.0 *Contributors:* Own work, based on :Image:Translation_arrow.svg. Created in Adobe Illustrator CS3 *Original artist:* tkgd2007

- **File:Ultrasound_of_embryo_at_5_weeks,_colored.png** *Source:* https://upload.wikimedia.org/wikipedia/commons/0/06/Ultrasound_of_ embryo_at_5_weeks%2C_colored.png *License:* CC BY 2.5 *Contributors:* File:Embryo at 5 weeks.JPG by X.Compagnion *Original artist:* Derivative by Mikael Häggström

- **File:Vetebrateembryo.svg** *Source:* https://upload.wikimedia.org/wikipedia/commons/d/d5/Vetebrateembryo.svg *License:* CC BY-SA 3.0 *Contributors:* Own work *Original artist:* Jlesk

- **File:Views_of_a_Foetus_in_the_Womb_detail.jpg** *Source:* https://upload.wikimedia.org/wikipedia/commons/2/2a/Views_of_a_Foetus_ in_the_Womb_detail.jpg *License:* Public domain *Contributors:* ? *Original artist:* ?

40.2.3 Content license